The IEE

D0333014

Protection Against Fire

GUIDANCE NOTE  4

4

IEE Wiring Regulations

BS 7671 : 2001 Requirements for Electrical Installations

Including Amd No 1 : 2002

Published by: The Institution of Electrical Engineers, Savoy Place,
LONDON, United Kingdom. WC2R 0BL

Printed August 1992
Reprinted April 1993, with amendments
Reprinted February 1994, with minor amendments
2nd edition October 1995
3rd edition Nov 1998, incorporating BS 7671 : 1992 inc Amd No 2
4th edition Aug 2003, incorporating BS 7671 : 2001 inc Amd No 1

Copies may be obtained from:

The IEE,
PO Box 96, STEVENAGE,
United Kingdom, SG1 2SD

Tel: +44 (0)1438 767 328
Fax: +44 (0)1438 742 792
Email: sales@iee.org
http://www.iee.org/Publish/Books/WireAssoc/

**While the author, publisher and contributors believe
that the information and guidance given in this work
is correct, all parties must rely upon their own skill
and judgement when making use of it. Neither the
author, the publisher nor any contributor assume any
liability to anyone for any loss or damage caused by
any error or omission in the work, whether such error
or omission is the result of negligence or any other
cause. Where reference is made to legislation it is not
to be considered as legal advice. Any and all such
liability is disclaimed.**

ISBN 0 85296 992 9, 2003

Contents

Co-operating Organisations

The Institution of Electrical Engineers acknowledges the contribution made by the following organisations in the preparation of this Guidance Note.

British Cables Association
    J M R Hagger BTech(Hons) AMIMMM

British Electrotechnical & Allied Manufacturers Association Ltd
    R Lewington MIEE

BRE Environmental
    K Bromley BA(Hons) MIEE

British Standards Institution
    W E Fancourt

CIBSE
    Eur Ing G Stokes BSc(Hons) CEng FIEE FCIBSE

City & Guilds of London Institute
    H R Lovegrove IEng FIIE

Electrical Contractors' Association
    D Locke IEng MIIE ACIBSE

Electrical Contractors' Association of Scotland t/a SELECT
    D Millar IEng MIIE MILE

Electricity Association Limited
    D J Start BSc CEng MIEE

Engineering Equipment and Materials Users Association
    R J Richman

EIEMA
    Eur Ing M H Mullins BA CEng FIEE FIIE

ERA Technology Ltd
    M W Coates BEng

Health & Safety Executive
    Eur Ing J A McLean BSc(Hons) CEng FIEE FIOSH

Institution of Electrical Engineers
    G D Cronshaw IEng MIIE (Editor)
    P R L Cook CEng FIEE MCIBSE
    P E Donnachie BSc CEng FIEE
    D W M Latimer MA(Cantab) CEng FIEE
    B J Lewis BSc Mphil CEng FIEE
    J Simmons CEng MIEE

National Inspection Council for Electrical Installation Contracting

Office of the Deputy Prime Minister
    E N King BSc CEng FCIBSE

Royal Institute of British Architects
    J Reed ARIBA

Acknowledgements

References to British Standards are made with the kind permission of BSI. Complete copies can be obtained by post from:

BSI Customer Services
389 Chiswick High Road
London   W4 4AL

Tel:	General Switchboard:	020 8996 9000
	For ordering:	020 8996 9001
	For information or advice:	020 8996 7111
	For membership:	020 8996 7002
Fax:	For orders:	020 8996 7001
	For information or advice:	020 8996 7048

BSI operates an export advisory service - Technical Help to Exporters - which can advise on the requirements of foreign laws and standards. The BSI also maintains stocks of international and foreign standards, with many English translations.
Up-to-date information on BSI standards can be obtained from the BSI website
http://www.bsi-global.com/

Extracts from Approved Document B, Fire Safety, Guidance on the Building Regulations 2000 edition, are reproduced with the kind permission of the Stationery Office. Complete copies can be obtained by post from the Stationery Office,
St Crispin's, Duke Street, Norwich, NR3 1PD.
Telephone orders: 01603 622 211.
Fax orders: 0870 600 5533.

Complete copies of Approved Document B are also downloadable from the Office of the Deputy Prime Minister's website
http://www.safety.odpm.gov.uk/bregs/brads.htm

Extracts from the Building Standards (Scotland) Regulations 1990 as amended are reproduced with the kind permission of the Stationery Office. Copies

may be obtained from the Stationery Office
Bookshop, 71 Lothian Road, Edinburgh, EH3 9AZ.
Telephone orders: 0870 606 5566.
Fax orders: 0870 606 5588.

Complete copies are also downloadable from the
Scottish Executive website
http://www.scotland.gov.uk/build_regs/standards/contents.asp

Preface

This Guidance Note is part of a series issued by the Wiring Regulations Policy Committee of the Institution of Electrical Engineers to enlarge upon and simplify some of the requirements of BS 7671 : 2001 inc Amd No 1, Requirements for Electrical Installations (IEE Wiring Regulations Sixteenth Edition). Significant changes made in this 4th edition of the Guidance Note are sidelined.

Note that this Guidance Note does not ensure compliance with BS 7671. It is a guide to some of the requirements of BS 7671 but users of these Guidance Notes should always consult BS 7671 to satisfy themselves of compliance.

The scope generally follows that of the Regulations and the principal Section numbers are shown on the left. The relevant Regulations and Appendices are noted in the right-hand margin. Some Guidance Notes also contain material not included in BS 7671 but which was included in earlier editions of the Wiring Regulations. All of the Guidance Notes contain references to other relevant sources of information.

Electrical installations in the United Kingdom which comply with BS 7671 are likely to satisfy the relevant aspects of Statutory Regulations such as the Electricity at Work Regulations 1989, but this cannot be guaranteed. It is stressed that it is essential to establish which Statutory and other Regulations apply and to install accordingly. For example, an installation in premises subject to licensing may have requirements different from, or additional to, BS 7671, which will take precedence.

Users of this Guidance Note should assure themselves that they have complied with any legislation that post-dates the publication.

Introduction

This Guidance Note is concerned primarily with Chapter 42 — Protection against thermal effects, and Chapter 48 — choice of protective measures as a function of external influences. It does not attempt to deal comprehensively with the safety of personnel in the event of fire, since this is beyond the scope of BS 7671.

Neither BS 7671 nor the Guidance Notes are design guides. It is essential to prepare a full design and specification prior to commencement or alteration of an electrical installation. Compliance with the relevant standards should be required.

The design and specification should set out the requirements and provide sufficient information to enable competent persons to carry out the installation and to commission it. The specification must include a description of how the system is to operate and all the design and operational parameters. It must provide for all the commissioning procedures that will be required and for the provision of adequate information to the user. This will be by means of an operational and maintenance manual or schedule, and 'as fitted' drawings if necessary.

514-09

It must be noted that it is a matter of contract as to which person or organisation is responsible for the production of the parts of the design, specification construction and verification of the installation and any operational information.

The persons or organisations who may be concerned in the preparation of the works include:

The Designer
The Planning Supervisor
The Installer
The Supplier of Electricity
The Installation Owner and/or User

The Architect
The Fire Prevention Officer
All Regulatory Authorities
Any Licensing Authority
The Health and Safety Executive

In producing the design, advice should be sought from the installation owner and/or user as to the intended use. Often, as in a speculative building, the intended use is unknown. The specification and/or the operational manual must set out the basis of use for which the installation is suitable. | 131-01-01

Precise details of each item of equipment should be obtained from the manufacturer and/or supplier and compliance with appropriate standards confirmed. | 511

The operational manual must include a description of how the system as installed is to operate and all commissioning records. The manual should also include manufacturers' technical data for all items of switchgear, luminaires, accessories, etc and any special instructions that may be needed. The Health and Safety at Work etc Act 1974 Section 6 and the Construction (Design and Management) Regulations 1994 are concerned with the provision of information, and guidance on the preparation of technical manuals is given in BS 4884 (Technical manuals. Specification for presentation of essential information, guide to content and guide to presentation) and BS 4940 | 514-09
(Recommendations for the presentation of technical information about products and services in the construction industry). The size and complexity of the installation will dictate the nature and extent of the manual.

Section 1 — The Statutory Requirements

General

A number of enactments and statutory instruments including the Electricity at Work Regulations 1989 (made under the Health and Safety at Work etc Act 1974) deal with fire risks, prevention of fire and fire precautions.

This Guidance Note is not intended to provide an exhaustive treatment of the legislation concerned with fire, but deals only with those situations referred to in BS 7671. Thus, certain specialised installations listed in Regulation 110-02 are excluded.

110-02

1.1 Statutory Regulations 13

1.1.1 *The Electricity at Work Regulations 1989*

Appx 2

The Electricity at Work Regulations are general in their application and refer throughout to 'danger' and 'injury'. Danger is defined as risk of 'injury' and injury is defined in terms of certain classes of potential harm to persons. Injury is stated to mean death or injury to persons from:

— electric shock

— electric burn

— electrical explosion or arcing

— fire or explosion initiated by electrical energy.

The Memorandum of Guidance on the Electricity at Work Regulations 1989 (HSE Publication HSR25) is essential reading for all concerned with electrical installations.

1.1.2 *The Electricity Safety, Quality and Continuity Regulations 2002*

The definition of 'danger' in these Regulations includes 'danger to health or danger to life or limb from electric shock, burn, injury or mechanical movement to persons, livestock or domestic animals, or from fire or explosion, attendant upon the generation, transmission, transformation, distribution or use of energy'.

The Regulations were made on the 24th October 2002, laid before Parliament 28th October 2002, and came into force on 31st January 2003.

1.1.3 *The Management of Health and Safety at Work Regulations 1999*

These Regulations place general duties on employers to assess risks to the health and safety of employees and others and take managerial action to minimise these risks, including:
— implementing preventive measures
— providing health surveillance
— appointing competent people
— setting up of procedures
— provision of information
— training etc.

1.1.4 *The Provision and Use of Work Equipment Regulations 1998*

These Regulations require employers to ensure that work equipment is suitable for the purpose.

The Regulations also require that work equipment is maintained in an efficient state, in efficient working order and in good repair.

1.1.5 *The Construction (Design and Management) Regulations 1994*

The Construction (Design and Management) Regulations (CDM Regulations) require active planning, co-ordination and management of the building works, including the electrical installation, to ensure that hazards associated with the construction, maintenance and perhaps demolition of the installation are given due consideration, as well as provision for safety in normal use.

1.1.6 *The Dangerous Substances and Explosive Atmospheres Regulations 2002*

These Regulations were made on the 7th November 2002, laid before Parliament 15th November 2002, came into force (in part) on the 9th December 2002 and fully on 30th June 2003.

1.1.7 *The Highly Flammable Liquids and Liquefied Petroleum Gases Regulations 1972*

Appx 2

These Regulations have particular applications where such substances are present for the purposes of, or in connection with, any undertaking, trade or business.

The Regulations will be repealed six months after the coming into force of the Dangerous Substances and Explosive Atmospheres Regulations 2002 (DSEAR 2002).

1.1.8 *The Petroleum (Consolidation) Act 1928*

Appx 2

This Act has particular application to any place where petroleum spirit is kept or handled.

Guidance on the design, construction, modification and maintenance of petrol filling stations is published by the Association for the Petroleum and Explosives Administration (APEA) and the Institute of Petroleum (IP), ISBN 0 85293 217 0. Copies of this document can be obtained from Portland Press Ltd, Commerce Way, Colchester, CO2 8HP. Tel 01206 796351.

The Health and Safety Executive publish two leaflets, one on the safe use of petrol in garages and one on dispensing petrol as a fluid. HSE Publication HSG41 Petrol filling stations: Construction and operation is currently out of print but remains a valued reference document, particularly with reference to the operation of petrol filling stations.

1.1.9 *Cinematography (Safety) Regulations 1955*

These Regulations apply to premises used for cinematography exhibitions.

Appx 2

1.1.10 *Building Regulations 2000*

Approved Document B 2000, approved by the Office of the Deputy Prime Minister as practical guidance on meeting the Requirements of Schedule 1 (means of escape) and Regulation 7 (materials and workmanship) of the Building Regulations.

See Appendix B of this Guidance Note for further information.

1.1.11 *Building Standards (Scotland) Regulations 1990 as amended*

The Scottish Office have produced Technical Standards to support the Building Standards (Scotland) Regulations 1990. The standards have full statutory force in Scotland by virtue of Regulation 9.

Reference is made to a number of British Standards which as a consequence of Regulation 9 have full statutory force in the particular circumstances e.g. BS 7671, CP 1007 : 1955, BS 5266 Part 1 : 1999, BS 5446 Part 1 : 2000.

See Appendix C of this Guidance Note for further information.

1.1.12 *The Health and Safety (Safety Signs and Signals) Regulations 1996*

These Regulations require employers to use a safety sign where there is a significant risk to health that cannot be controlled or avoided by other means. The HSE guide, Safety Signs and Signals (L64), describes in detail a range of signs including Emergency Escape and Fire Safety Signs.

1.1.13 *The Fire Precautions (Work Place) Regulations 1997*

These Regulations require employers where necessary, to safeguard the safety of employees, in case of fire. They include requirements to equip the work place with fire-fighting equipment and with fire detectors and alarms, as necessary. Guidance is provided on the Regulations in Home Office and HSE Publication "Fire Safety: an Employers Guide" - ISBN 0 11341 2290" and "Guide to fire precautions in existing places of entertainment and like premises" - ISBN 0 11340 9079.

Electrical installations are subject to many local authority bylaws and conditions of license. The following may be mentioned:

— places licensed for public entertainment, music, dancing, etc

— large buildings, office blocks and the like, having limited access and escape points

— buildings to which the public have access, covered shopping centres and the like, hotels and boarding-houses.

Section 20 of the London Buildings Acts (Amendment) Act 1939 (as amended primarily by the Building (Inner London) Regulations 1985) is principally concerned with the danger arising from fire within certain classes of buildings which, by reason of height, cubical extent and/or use necessitate special consideration. The types of building coming within these categories are defined under Section 20 of the amended 1939 Act.

The designer of the electrical installation in any building falling within these descriptions should make the earliest possible approach to the Local Authority and the Local Fire Authority to establish their requirements, obtain advice and get written agreement to the proposed arrangements.

Section 2 — The Wiring Regulations

General
13

Chapter 13 of BS 7671 prescribes the fundamental principles for safety and includes a number of regulations which are directly concerned with the fire risk associated with electrical installations.

Chap 13

In particular, Regulation 130-03-01 deals with the risk of ignition of flammable materials due to high temperature or electric arc and Regulation 130-03-02 deals with the effects of heat or thermal radiation emitted by electrical equipment. Other Chapters set out in greater detail the methods and practices which are regarded as meeting the requirements of Chapter 13.

130-03-01

130-03-02

For example, Chapter 42, 'Protection against thermal effects', has requirements for protection against fire, harmful thermal effects and burns. Also Chapter 43 prescribes requirements which limit the temperatures of live conductors and their insulation to prevent damage to the cable under overload and fault conditions.

Chap 42

433-01-01
434-01-01

Chapter 48, 'Choice of protective measures as a function of external influences', contains requirements for precautions to be taken where particular risks of the danger of fire exist. The Chapter contains the requirements to be met to prevent the dangers of fire and applies in locations with risks of fire due to the nature of processed or stored materials, and installations in locations constructed of combustible materials.

Chap 48

Chapter 54 has the same intent as Chapter 43 as regards protective conductors under earth fault conditions. Both these Chapters are the subject of Guidance Note 6: Protection Against Overcurrent.

Chap 54

GN6

It should be noted that BS 7671 does not contain specific requirements for electrical installations in areas which may be designated as escape routes. However, concern is often expressed over the impairment of escape routes by the smoke and fumes

131-05-02

generated by fire. This is an important but very complex issue and the risk from electrotechnical products should not be considered separately but as part of an overall fire hazard assessment.

Where consideration of the effects of smoke, fumes, fire propagation and/or circuit integrity is necessary, cables which have characteristics classified in standard fire tests can be used. The cable standards are:

BS 6724 : 1997 'Specification for 600/1000 V and 1900/3300 V armoured electric cables having thermosetting insulation and low emission of smoke and corrosive gases when affected by fire'.

BS 6724

BS 7211 : 1998 'Thermosetting insulated cables (non-armoured) for electric power and lighting with low emission of smoke and corrosive gases when affected by fire'.

BS 7211

BS 7629-1 : 1997 'Specification for 300/500 V fire-resistant electric cables having low emission of smoke and corrosive gases when affected by fire. Multicore cables'.

BS 7629-1

BS 7846 : 2000 'Electric cables, 600/1000 V armoured fire-resistant electric cables having thermosetting insulating and low emission of smoke and corrosive gases when affected by fire'.

BS 7846

Cables clipped direct may also be mineral insulated to BS EN 60702-1, either bare or, when a covering is required, it should be designated as having 'low emission of smoke and corrosive gases when affected by fire'.

BS EN 60702-1

General guidance on the selection of cables is given in BS 7540 : 1994 'Guide to the use of cables with a rated voltage not exceeding 450/750 V'. For 600/1000 V cables guidance is contained in the individual product standards.

BS 7540

Section 605, Agricultural and horticultural premises, includes a requirement for the installation of an RCD having a rated residual operating current not exceeding 500 mA for protection against fire in some specified situations. It may be appropriate to consider the use of such an RCD in other situations where the possibility of fire due to an earth fault is of particular concern. Chapter 48 mentioned earlier contains a

605-10-01

482-06-02

similar requirement, though this prescribes an RCD having a rated residual current not exceeding 300 mA, which is more likely to be used than a 500 mA device.

It may be necessary to consider whether, in particularly vulnerable situations where readily ignitable materials are present (Section 5 of this Guidance Note), an RCD rating of less than 500 mA or 300 mA would be appropriate for protection against fire.

Section 3 — Thermal Effects

General
120-01

The primary purpose of BS 7671 is to ensure that persons, property and livestock are protected against hazards arising from the electrical installation. The particular risks identified are:

— shock
— burns, fire and other injurious effects
— explosion
— mechanical injury.

130-01-01

3.1 Thermal effects 421

Thermal effects (fire and burns) associated with an electrical installation may be categorised as follows:

421-01-01

   (i) the ignition hazard arising directly from the installation, and

422

   (ii) the fire propagation hazard where the components of the installation contribute to the risk, initiated from whatever source, and

527

   (iii) the risk of burns from electrical equipment, and

423

   (iv) the impairment of the safe working of equipment, and

   (v) external influences.

3.2 The ignition hazard 422

General

Under this heading fixed equipment, joints, connections and flammable liquids within electrical equipment are considered.

3.2.1 Fixed equipment 422

Many fires attributed to electricity are caused by portable equipment. Neither the installation designer nor the installer has any control over the selection and use of such equipment. Chapters 42 and 48 are concerned solely with fixed equipment.

Section 482 sets out the precautions necessary in locations where a particular danger of fire exists. These are locations with risks of fire due to the nature of processed or stored materials and locations with combustible constructional materials.

482

Section 422 sets out the precautions necessary to prevent the heat generated by equipment creating the risk of fire, or of harmful thermal effects to adjacent fixed material. BS 7671 also places a duty on the installation designer to exercise foresight and consider other material which is likely to be in proximity to such equipment. For example, it is foreseeable that kitchen windows would be curtained and this must influence the positioning of an electric cooker. In locations intended for livestock, suitable measures must be taken to avoid burns which might lead to panic.

422-01-01

605-10-02

BS 7671 also refers to compliance with any relevant installation instruction of the equipment manufacturer. This is because of the general requirement that equipment complies with a relevant standard. Many British Standards such as BS EN 60335-1 (currently partially replaced) for household appliances, BS EN 60950 for information technology equipment and office machines and BS 60335-2-36 : 1995 for commercial catering equipment include a test to evaluate the thermal effect on the surroundings.

422-01-01

511

BS EN 60335-1
BS EN 60950

BS 60335-2-36

When an installation includes fixed equipment, the fact that the selection of the equipment may have been made by other than the installer does not absolve the installer from the responsibility of seeing that any installation instructions of the manufacturer are met.

Regulation 422-01-02 offers three installation methods for equipment which in normal operation has a surface temperature sufficient to cause a risk of fire or harmful effects to adjacent materials.

422-01-02

One or more of the methods must be adopted and as far as Regulation 422-01-02(iii) is concerned, Fig 3.1 gives guidance for equipment which in normal operation has a surface temperature exceeding 90 °C.

Fig 3.1: Elevation

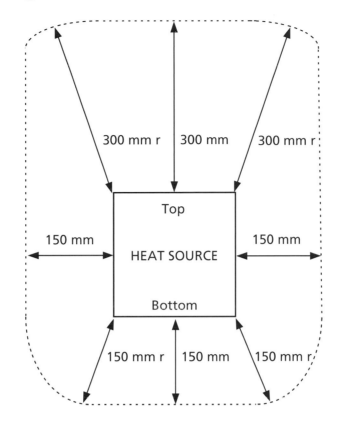

By definition, the term 'equipment' includes luminaires and the heating effect of these must be considered.

422-01-01

Particular attention must be given to recessed or semi-recessed luminaires and luminaire controlgear mounted in ceiling voids to ensure that heat is properly dissipated and that thermal insulating materials are not and cannot become so disposed as to restrict the cooling of the equipment.

Lampholders are one example of equipment having temperature ratings. BS EN 61184 : 1997 'bayonet lampholders' gives guidance on the selection of lampholders for particular applications. However, because it is not possible to envisage how an installation will be used, every B15 or B22 lampholder must be T2 temperature rated to comply with BS 7671. Where lampholders other than B15 or B22 are installed, their rated operating temperature must be suitable for the application.

BS EN 61184

553-03-03

The surface temperatures attained by electrical equipment are also the subject of Statutory

Regulations. For example, the Highly Flammable Liquids and Liquefied Petroleum Gases Regulations, 1972 require that if a process involves cellulose nitrate e.g. spray painting, then the temperature of any surface upon which residues may be deposited shall not exceed 120 °C. Alternatively, guards may be provided to prevent such deposition. These Regulations will be repealed six months after the coming into force of the Dangerous Substances and Explosive Atmospheres Regulations 2002. See paragraph 1.1.6 of this Guidance Note.

From the above it will be apparent that where potentially dangerous substances are involved there are probably statutory requirements to be complied with. A thorough study of the legislation, as well as the relevant British Standards, followed by discussion and agreement with the Health and Safety Executive and the Fire Authority is necessary before a design can be finalised or work undertaken (see also Section 5).

3.2.2 Arcing or the emission of high temperature particles 422

The effects of arcs or the emission of high temperature particles must be guarded against. In some equipment, e.g. air circuit-breakers or semi-enclosed fuses, emissions may be produced during fault clearance and the enclosures of such equipment must comply with the appropriate British Standard, or be screened by arc-resistant material, or be mounted so as to allow for the safe extinction of the emissions at a sufficient distance from the material upon which the emissions could have a harmful effect.

422-01-03

511

By its application to any item of fixed equipment, BS 7671 also applies to permanently installed electric welding sets. Electric arc welding and similar, fixed or portable, is subject to Regulations 7, 8 and 14 of The Electricity at Work Regulations 1989. Regulation 7 requires insulation of live conductors, Regulation 8 earthing, and Regulation 14 specifies requirements for work on or near live conductors. Further guidance is given in HSE Publication 'PM64 — Electrical safety in arc welding'.

Electricity at Work Regulations

3.2.3 The ignition hazard — joints and connections

422

Regulation 422-01-04 is the general requirement for the enclosure of terminations of live conductors and joints. Reference is made to Regulation 526-03 where particular requirements are given, including also the enclosure of terminations of PEN conductors. (See also paragraph 3.3.1 below.) The first requirement is to provide an enclosure which will protect the joint or connection against the environment.

422-01-04

526-03

Poor terminations and connections are a frequent cause of fire and close attention is required by the designer, the installer and the person responsible for the inspection and testing to all aspects of the subject.

422-01-04
526-01 to
526-04
712-01-03

For this reason, Regulation 526-03-02 requires that all terminations and joints at whatever voltage, including ELV, shall be made within a suitable enclosure. Because there is always a risk of overheating and consequent fire at a joint or termination these must be enclosed and the enclosure where not incorporated in suitable equipment or accessory must meet the specified fire resistance requirements. This applies equally to ELV connections to luminaires and similar equipment. The relatively high currents of ELV equipment means particular care has to be taken with joints and terminations. When compression joints are used, crimping tool, lug and cable must be compatible.

526-03-02

Other matters which should be considered under this heading are:

(i) dirty or misaligned equipment contacts which may give rise to local heating, and

(ii) loose or inadequate cable supports which may place mechanical stresses on connections causing overheating.

One method, which could be considered when inspecting electrical installations for signs of overheating, is the use of thermal imaging. This is not required by BS 7671 and would therefore be in addition to the Schedule of Inspections and tests stated in Part 7 of BS 7671. Thermal imaging of exposed live parts of electrical installations must comply with the requirements of Regulation 14 of the Electricity at Work Regulations 1989. Regulation 14 prohibits a person working on or near any exposed bare live conductor unless it is unreasonable in all circumstances for it to be dead, and it is reasonable in all circumstances for the person to be at work on or near it while it is live and suitable precautions taken to prevent injury.

3.2.4 Flammable dielectric liquids 422

Special precautions are necessary for flammable dielectric liquids, as fires involving these liquids are life threatening within a few seconds of ignition. In the event of spillage the object is to limit the spread of the liquid and the exposed surface area, thus limiting the size of the fire and the danger to persons and property.

422-01-05

Where a large quantity of liquid is involved more than one escape route is necessary, in the same way as for a large switchroom. It is important to ensure that persons have alternative routes so that the main seat of a fire may be avoided.

The 'single location' referred to in BS 7671 is the location containing all the flammable liquid which may be involved at the outset of an emergency. The amount of flammable liquid in an area must therefore be established, whether contained in one item of equipment or in a number of separate items.

422-01-05

The options available to the designer will depend on a number of considerations, for example, whether a single item of equipment is involved or a number of items and whether the location is indoors or outdoors. The options include :
— reducing the risk by partitioning the location with fire doors and sills
— providing bunds or kerbs around items of equipment or, for larger items of equipment, a catchpit filled with pebbles or granite chips (the net capacity of the bund or catchpit when filled with pebbles or chips should exceed the oil capacity of the equipment by at least 10%)
— providing a drainpit and flame arrestor
— provision of automatic fire venting and/or automatic fire suppression or foam inlets
— ramped floors
— use of an outdoor location
— blast walls between large items.

In some circumstances, it may be appropriate to provide explosion venting for interior locations and environmental considerations should always involve discussions with public health engineers concerning the provision of an oil interceptor to prevent contamination of the sewers.

When carrying out inspections of older installations the measures provided should be carefully examined in the light of modern practice. In particular, it may be that over a period of time, seepage and occasional spillage has caused a wooden floor to be saturated with oil and thus become an additional fire risk.

3.2.5 Focusing of heat

Equipment that focuses heat, such as radiant heaters and high intensity luminaires, must be positioned so that the building structure or other materials are not subjected to excessive temperatures; see also section 5.2. The manufacturer's instructions must be followed with respect to spacing from walls (w) floor (f) and ceiling (c) and the angle of inclination θ of the heater (Fig3.2). Regulation 482-02-13 states the *minimum* distances that spotlights and projectors must be installed from combustible materials.

422-01-06

482-02-13

Fig 3.2: Focusing of heat from a heater or luminaire

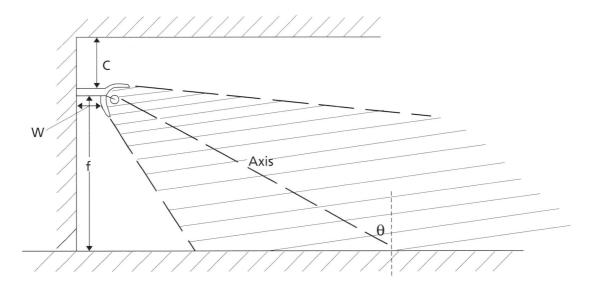

3.2.6 Enclosures

Where enclosures are constructed on site, reference must be made to the appropriate product standard for the necessary resistance to heat and fire. In the absence of such a standard, maximum likely temperatures of the enclosure must be determined

422-01-07

25

and enclosures selected that are well able to withstand these temperatures.

BS 476 Fire tests on building materials and structures, (Part 4 Non-combustible tests for materials, Part 12 Method of test for ignitability of products by direct flame impingement) provides type tests for materials.

When equipment is installed in an enclosure or enclosed space e.g. ballasts or other controlgear within a ceiling, adequate arrangements must exist for safely dissipating heat generated.

3.3 The propagation hazard 527

Under this heading are considered combustible components of an electrical installation which may contribute to the spread of fire through a building and also other matters concerned with fire aspects of the wiring system.

The contribution of electrical equipment including the cable system to the overall fire hazard cannot be considered separately from the whole building. These matters must be co-ordinated under the direction of the planning supervisor for the project. The Construction (Design and Management) Regulations are relevant.

3.3.1 Combustible components of the installation 527

Items such as cable insulation and wooden backboards for switchgear or consumer units are unlikely to be of significance in an already well-established fire.

However, what might be of considerable significance is that they may provide sufficient fuel to propagate a fire of electrical origin which, in their absence, would not spread beyond the immediate location.

Enclosure of non-sheathed cables in conduit, trunking or ducting may be a way of restricting the propagation of fire. If non-metallic trunking, ducting or conduit are to be used, non-flame propagating systems to BS EN 50085 or BS EN 50086 may be installed without further precautions. BS EN 50086 requires that flame propagating conduit be orange in colour.

521-07
Table 4A1
521-05
527-01-05
BS EN 50085
BS EN 50086
Clause 7.3

Similarly, terminations and joints of cables are required to be enclosed in a suitable accessory or enclosure. This may comply with the appropriate British Standard or be formed of material which has

422-01-04
526-03-02

passed the test referred to in Regulation 526-03-02(iv) and (v).

Cables for general use are required to comply with BS EN 50265-2-1, except for short lengths of cable connecting appliances to the fixed installation. 527-01-04

Although less common nowadays, backless consumer units are sometimes encountered. Such units should be fitted with a back of material considered to be non-combustible when tested to BS 476 Part 4 or be fixed directly to the building structure on material having the ignitability characteristic 'P' of BS 476 Part 5. 526-03-02(iv) 526-03-02(v) BS 476

3.3.2 Spread of fire 527

Selection and erection of all wiring systems must be made with the fire safety and integrity of the building in mind.

The sealing of wiring system penetrations is of the utmost importance. Making good with mortar or plaster might be appropriate for penetration of solid elements. Proprietary materials are also available for the purpose and include intumescent pastes and blocks and mineral fibres. The advice of the builder must be sought and followed. 527-02

For extreme conditions, patented enclosures are available.

An important exemption from the requirement for internal sealing is contained in Regulation 527-02-02, applying to non-flame propagating wiring systems having an internal cross-sectional area not exceeding 710 mm^2. This area covers standard 32 mm conduit, ducting or trunking of, say, 25 mm x 25 mm. For the purpose of this regulation non-flame propagating cable management systems are: 527-02-02

1. metallic conduit and trunking, and
2. conduit systems classified as non-flame propagating to BS EN 50086, and BS EN 50086
3. ducting and trunking systems classified as non-flame propagating to BS EN 50085 or BS 4678. BS EN 50085 BS 4678

Temporary sealing arrangements during the course of the work may be necessary and should always be considered during works of refurbishment or rebuilding. 527-03-01

Bags of intumescent material which are available in various sizes are particularly useful for this purpose.

Regulation 527-04-01 requires that each sealing arrangement be inspected while it is still accessible and the details recorded. Regulation 711-01-01 requires that every installation shall, during erection and on completion before being put into service, be inspected and tested to verify, so far as is reasonably practicable, that the requirements of the Regulations have been met. This should be regarded as part of the inspection procedure, Regulation 712-01-03(xviii) — erection methods.

527-04-01

711-01-01

712-01-03(xviii)

3.3.3 Spread of fire luminaire diffusers

There are provisions that apply to lighting diffusers which form part of a ceiling in Section 7 of Approved Document B, Fire Safety 2000 edition. This is issued by the Office of the Deputy Prime Minister and has been approved by the Secretary of State as practical guidance on meeting the requirements of Schedule 1 and Regulation 7 of the Building Regulations — see Appendix B. There is no restriction on the use of diffusers of classification TP(a), but diffusers of classification TP(b) are restricted as per Table 11 and Diagram 24 of Appendix B. (Definitions of classifications TP(a) and TP(b) are given in the notes to Appendix B.)

Faulty starters are reported as causing TP(b) diffusers to ignite, resulting in fires.

In Scotland the requirements of Technical Standard E6.1 need to be met.

3.4 The risk of burns 423

The requirements of Section 423 of BS 7671 concerning burns apply only to burns caused by contact with heated surfaces. Heat radiation or arc burns are considered to be covered by the measures required in Regulations 422-01-02 and 422-01-03.

423-01-01

422-01-02
422-01-03

It should be noted that micc cables exposed to touch are permitted to have a sheath temperature of 70 °C, corresponding to a metallic part intended to be touched but not hand-held. However, a cable having a conductor operating temperature of 90 °C may achieve a surface temperature approaching 80 °C in normal operation.

Table 42A

It must always be borne in mind that the temperatures of Table 42A are maximum values and that contact with any surface at or above 70 °C may cause a dangerous reflex action.

BS 4086 : 1966 (1995) 'Recommendations for maximum surface temperatures of heated domestic equipment' provides technical considerations and recommended maximum temperatures for controls and working surfaces of heated domestic equipment. BSI PD 6504 : 1983 'Medical information on human reaction to skin contact with hot surfaces', provides information prepared by medical experts on human reaction to contact with heated surfaces. Both provide good guidance in determining 'safe' surface temperatures.

BS 4086

PD 6504

Even if the equipment complies with its standard as to surface temperature, consideration must still be given to the risk of burns particularly when equipment is to be installed in locations to be used by the very young or infirm. Additional precautions may be necessary, such as guards over heaters.

3.5 The risk of overheating 424

Section 424 of BS 7671 prescribes requirements for forced air heating systems and appliances producing hot water or steam.

Electricaire heating systems are excluded from the requirements. Such systems must comply with BS EN 60335-2-77 : 2001 (BS 3456 Part 2 Section 2.22 electricaire heaters remains current).

BS EN 60335-2-77

Approved Document B 2000 edition Fire Safety, paragraph 2.15 lists the precautions needed to avoid the possibility of the system allowing smoke or fire to spread into a protected stairway. Guidance is set out in Clause 6 of BS 5588, Part 1. BS 5588 covers fire precautions in the design, construction and use of buildings. Part 1 is a 'Code of Practice for Residential Buildings'.

BS 5588

Storage water heaters should comply with BS EN 60335-2-21 : 1999. Unvented hot water storage units and packages must comply with BS 7206.

BS EN 60335-2-21
BS 7206

3.6 Selection and erection 522	There are other provisions of Chapter 52 that concern wiring systems and the risk of fire, some of which are discussed below.	Chap 52

3.6.1 Ambient temperature

Regulation 522-01: Ambient temperature, is relevant because it affects the current-carrying capacity of elements of the installation e.g. cables and switchgear. Correct selection of cables for current-carrying capacity is essential and due regard must be paid to derating where cables pass through or are in contact with thermal insulating material. Following on from this, Regulation 522-02 deals with methods of mitigating the effects of external heat sources.

522-01
522-02
523
523-01

523-04

522-02

3.6.2 Flexible cords

Regulation 522-02-02 is concerned with parts of a cable or flexible cord within an accessory, appliance or luminaire. Such cables or flexible cords must either be suitable for the temperatures likely to be encountered or be further protected with additional insulation such as high temperature resistant sleeving.

522-02-02

3.7 External influences

In dirty or dusty locations, the build-up of deposits on conduits, cables, cable trays, etc cannot be ignored. In extreme circumstances, the heat dissipation from the wiring system could be affected and perhaps the dust itself could be a source of fire or explosion risk. The designer and installer should be aware of these dangers and adopt suitable measures. Such dusts include:

522-04-02

— sugar
— flour
— wood dust
— textile 'fly'
— starch
— spray paint
— coal.

Compliance with BS 7671 only may not be sufficient in such circumstances.

It may be appropriate to use equipment complying with BS EN 50281-1-1 and -2 : 1999 (BS 6467 remains current) 'Electrical apparatus with protection by enclosure for use in the presence of combustible dusts'. Part 1 covers the specification for apparatus

BS EN 50281-1-1
BS EN 50281-1-2

and Part 2 is a guide to selection, installation and maintenance.

Good housekeeping is essential particularly in dusty environments. Programmed maintenance including cleaning is likely to be required.

Section 4 — Use of the Installation and Alterations and Additions

General 73

Almost every installation is subject to some abuse during its lifetime and, in addition, alterations and additions are made which may not accord with the intentions of the original designer and installer. These are matters which demand attention during any periodic inspection.

731
732

4.1 The use of the installation 523

Among the causes of increased fire risk due to unconsidered actions are:

(i) cables originally correctly selected, but subsequently overloaded because of a change in the connected load

523
Appx 4

(ii) cables which have been surrounded by thermal insulation or other materials after installation

523-04

(iii) conduit, trunking and ducting originally correctly sized for cable capacity, but advantage being subsequently taken of vacant space to install extra circuits causing overheating of all the cables

Appx 4

(iv) steam pipes (either for heating or process purposes) inadequately lagged and near to cables or trunking

528

(v) the work carried on in premises, or part of them, served by an installation may have changed. For example, adverse conditions of dust may have been introduced, there may be corrosive fumes not previously present, or flammable materials may be processed or stored in an area previously free of such activities (see paragraph 5.1)

528

(vi) motors or other heat-producing equipment may be deprived of ventilation by build-up of dust, which in itself may be flammable, or by other obstructions (see paragraphs 3.7 and 5.1)

BS EN 50281-1-1
BS EN 50281-1-2

(vii) wiring not considered liable to mechanical damage, for instance in a relatively inaccessible roof space, may be damaged by future activities such as plumbing or heating alterations, roof repairs, or the storage of heavy or sharp-edged articles

522-06

(viii) reflector spot lamps e.g. for curing, fitted in luminaires not designed to accept them, causing a concentration of heat either in front of or behind the lamp

422-01-01

(ix) missing joint box covers or unplugged entries may invite vermin damage to cables, creating the risk of subsequent arcing or of faults caused by vermin coming into contact with live conductors.

4.2 Alterations and additions 721

When alterations or additions to an installation are made, care should be taken as far as is reasonably practicable to remove all redundant materials which may be a source of fire propagation.

In particular, cables, whether enclosed or unenclosed, should be removed together with wooden items such as pattresses, backboards, etc. and any packaging material.

4.3 Periodic inspections 731

Items which may give rise to risk of fire in electrical installations and which should also be included in any periodic inspection are:

GN3
Inspection & Testing

(i) frayed, worn or unsuitable flexible cords, perhaps with taped joints

(ii) flexible cords inadequately secured in cord grips

(iii) cables and flexible cords run under carpets, through doorways or in other vulnerable locations

(iv) bell wire used for mains voltage applications

(v) dangerous proliferation of adaptors (see also paragraph 8.3, Extension leads and adaptors)

(vi) signs of overheating at appliance and other connectors, joints and terminations generally

(vii) missing guards on heaters

(viii) fabric or plastic shades unsuitable for lamp wattage or type

(ix) heat sources including lamps so located that nearby fittings or furnishings could become so disposed as to be at risk of fire

(x) inadequate enclosures of cable joints and terminations.

This list is not exhaustive. Regular periodic inspections are necessary at a frequency that is appropriate to the premises and their use. Some guidance is given in Guidance Note 3, Inspection & Testing.

Section 5 — Locations with Increased Risk

5.1 Potentially explosive atmospheres, etc
11

Installations in potentially explosive atmospheres or in situations of unusual fire risk are included as item (xviii) of Regulation 110-01-01. This does not mean that such installations are outside the scope of BS 7671. It is only those aspects concerning the assessment of the degree of risk and the precautions which need to be taken which BS 7671 does not cover. BS 7671 provides assistance in the selection of cable sizes, methods of protection and installation, but it is necessary also to refer to the other relevant British Standards:

110-01-01(xviii)

(i) BS EN 60079 : 'Electrical apparatus for explosive gas atmospheres'

BS EN 60079

(ii) BS EN 50014 : 'Electrical apparatus for potentially explosive atmospheres'
Note: This replaced BS 5501 which has been withdrawn.

BS EN 50014

(iii) BS 6467 : 'Electrical apparatus with protection by enclosure for use in the presence of combustible dusts'.

BS 6467

Particular competence and appropriate industry training is required when designing and installing in such locations. Refer to paragraphs 1.1.6, 1.1.7 and 3.7.

Guidance on the design, construction, modification and maintenance of petrol filling stations is published by the Association for the Petroleum and Explosives Administration (APEA) and the Institute of Petroleum (IP), ISBN 0 85293 217 0. Copies of this document can be obtained from Portland Press Ltd, Commerce Way, Colchester, CO2 8HP. Tel 01206 796351.

The Health and Safety Executive publish two leaflets, one on the safe use of petrol in garages and one on dispensing petrol as a fuel. HSE Publication HSG41 Petrol filling stations: Construction and operation is currently out of print but remains a valid reference document, particularly with reference to the operation of petrol filling stations. In addition, there are proposals for new petrol legislation phase 1— Changes to workplace controls, and proposals for the Dangerous Substances and Explosive Atmospheres Regulations by the Health and Safety Commission.

5.2 Risks from materials

5.2.1 Risks from the process or stored materials

A clear distinction has to be made between installations in locations with explosion risks, which are discussed in paragraph 5.1 of this guide, and the locations covered in Chapter 48 of BS 7671 which deals with risks of fire due to the nature of processed or stored materials and locations with combustible constructional materials.

If the risk is from the process or stored materials (e.g. the manufacturing, processing or storage of combustible materials, the accumulation of materials such as dust and fibres in barns, woodworking factories, paper mills, textile factories or similar) and there is no explosion risk, Regulation 482-02 prescribes the following requirements:

(i) restrict electrical equipment to only that necessary for the location

482-02-01

(ii) where material such as dust or fibres , sufficient to cause a fire hazard, could accumulate on enclosures of electrical equipment , adequate measures shall be taken to prevent the enclosures from exceeding 90 °C under normal conditions and 115 °C under fault conditions

482-02-02

(iii) select only equipment suitable for its location with an enclosure that has an IP rating of at least IP 5X. The degree of protection provided by an enclosure is indicated by two numerals. The first indicates the degree of protection against solid bodies and the second indicates the degree of protection against liquids. Where a characteristic numeral is not required to be specified it can be replaced by the letter X. (Refer to IEE Guidance Note 1, Appendix B, for full details). In this case the first numeral 5 indicates protection against contact by tools, wires or strips more than 1.0 mm

482-02-03

thick, and dust-protected (dust may enter but not in an amount sufficient to interfere with satisfactory operation or impair safety).

(iv) A cable not completely embedded in non-combustible material such as plaster or concrete or otherwise protected from fire, shall meet the flame propagation characteristics as specified in BS EN 50265-2.1 or 2.2.

Conduit systems shall be in accordance with BS EN 50086 and ducting and trunking systems shall be in accordance with BS EN 50085. All such systems shall be non-flame propagating and comply with the resistance to flame propagation, resistance to abnormal heat and resistance to heat tests of BS EN 50085 or BS EN 50086. Flame propagating systems shall only be used where they are embedded in the building structure e.g. screeded systems.   *482-02-04*

Where the risk of flame propagation is high, e.g. in long vertical runs or bunched cables, the cable shall meet the flame propagation characteristics as specified in one of the categories of test in Parts 2.1 to 2.5 of BS EN 50266 (these have replaced BS 4066-3).

(v) A wiring system which passes through but is not intended for electrical supply within the location shall have an IP rating of IP 5X or meet the requirements in (iv) above as applicable, and have no joint within the location unless it is placed in a suitable enclosure that meets the flame propagation characteristics of the wiring system, such that it does not affect the flame propagation characteristics of the wiring system.   *482-02-05*

(vi) except for mineral insulated cables complying with BS EN 60702-1 and busbar trunking systems complying with BS EN 60439-2, wiring systems shall be protected against insulation faults to earth as follows: *482-02-06*

(i) in TN and TT systems, by an RCD with a rated residual operating current not exceeding 300 mA

(ii) in IT systems, by insulation monitoring devices with audible and visual signals. Adequate supervision is required to facilitate manual disconnection as soon as appropriate. In the event of a second fault, the disconnection time of the overcurrent protective device shall not exceed 5 s

(vii) PEN conductors shall not be used. This regulation does not apply to wiring systems which pass through the location   482-02-07

(viii) switch the neutral conductor as well as the phase conductors when isolating equipment. A linked switch or linked circuit-breaker is required   482-02-08

(ix) exposed bare and live conductors shall not be used   482-02-09

(x) flexible cables and flexible cords shall be of heavy duty type having a voltage rating of not less than 450/750 V or suitably protected against mechanical damage   482-02-10

(xi) protect all motors (which are automatically or remotely controlled or which are not continuously supervised) against excessive temperature by an overload protective device with manual reset. Motors with star-delta starting shall be protected against excessive temperature in both the star and delta connections. Also, it is advisable to protect motors with slipring starters from being left with the resistance in the rotor. This type of rotor has a 3-phase winding, the ends of which are connected to three sliprings on the rotor shaft. For this reason it is sometimes called a 'slipring' motor. This enables an external resistance to be added to the rotor circuit, which is used to   482-02-11

  a) limit the starting current

  b) give a high starting torque

  c) provide speed control.

At normal speed the sliprings are short-circuited and the brush gear is lifted clear of the sliprings to reduce wear

(xii) use only luminaires with limited surface temperatures. Fittings manufactured to BS EN 60598-2-24 have horizontal surface temperatures limited to 90 °C under normal conditions and 115 °C under fault conditions.

482-02-12

BS EN 60598

(xiii) except as otherwise recommended by the manufacturer, spotlights and projectors shall be installed at the following minimum distances from combustible materials:

482-02-13

   (i) rating up to 100 W – 0.5 m

   (ii) rating over 100 W up to 300 W – 0.8 m

   (iii) rating over 300 W up to 500 W – 1.0 m

(xiv) luminaires shall be of a type that prevents lamp components from falling from the luminaires. These are components that are likely to run hot, such as lamps

482-02-14

(xv) where heating and ventilation systems containing heating elements are installed, the dust or fibre content and the temperature of the air shall not present a fire hazard. Drawing air from areas where no dust is likely to be present is advisable. The temperature of the heated air must be limited. Temperature limiting devices according to Regulation 424-01 shall have manual reset. Regulation 424-01-01 requires two temperature limiting devices independent of each other which prevent permissible temperatures from being exceeded

482-02-15

424-01-01

(xvi) heating appliances shall be fixed. Where a heating appliance is mounted close to combustible materials, barriers shall be provided to prevent the ignition of such materials

482-02-16

(xvii) heating storage appliances shall be of a type which prevent the ignition of combustible materials and/or fibres by the heat storing core

482-02-17

(xviii) Enclosures of equipment such as heaters and resistors shall not attain higher surface temperatures then 90 °C under normal conditions, and 115 °C under fault conditions.

482-02-18

5.2.2 Risks from combustible constructional materials

If the risk is from locations with combustible constructional materials and there is no explosion risk Regulation 482-03 has the following requirements:

(i) electrical equipment, e.g. installation boxes and distribution boards, which is installed on or in a combustible wall shall comply with the relevant standards for enclosure temperature rise

482-03-01

(ii) where electrical equipment does not comply with the relevant standards for enclosure temperature rise, the equipment shall be enclosed with a suitable thickness of non-flammable material. The effect of the material on the heat dissipation from electrical equipment shall be taken into account

482-03-02

(iii) cables and cords shall comply with the requirements of BS EN 50265-2-1 or -2-2. These British Standards give test methods for cables for resistance to vertical flame propagation (flame spread) using a 1 kW pre-mixed flame and a diffusion flame respectively

482-03-03

(iv) conduit and trunking systems shall be in accordance with BS EN 50086-1 and BS EN 50085-1 respectively, and shall meet the fire tests within the standards. BS EN 50086-1 deals with resistance to flame propagation and heat and BS EN 50085-1 deals with resistance to flame propagation, heat and abnormal heat.

482-03-04

5.3 Smoke and corrosive gases

Cables with low emission of smoke and corrosive gases must be considered for use where such properties will reduce the risk of danger to persons and livestock and/or damage to property and equipment. Persons are particularly at risk where large numbers of people gather and/or where there may be escape problems, e.g. tube stations, public buildings, hospitals and old people's homes.

Low corrosion properties should be considered for such areas as computer suites and control rooms where fire in cabling may otherwise cause considerable damage.

The relevant cable standards are:

BS EN 60702-1 : 2002 'Mineral insulated cables and their terminations with a rated voltage not exceeding 750 V – Part 1: Cables'.

BS EN 60702-1

BS 6724 : 1997 'Specification for 600/1000 V and 1900/3300 V armoured electric cables having thermosetting insulation with low emission of smoke and corrosive gases when affected by fire'.

BS 6724

BS 7211 : 1998 'Specification for thermosetting insulated cables (non-armoured) for electric power and lighting with low emission of smoke and corrosive gases when affected by fire'.

BS 7211

BS 7629-1 : 1997 'Specification for 300/500 V fire resistant electric cables having low emission of smoke and corrosive gases when affected by fire. Part 1 : Multicore cables'.

BS 7629

BS 7846 : 2000 'Electric cables 600/1000 V armoured fire resistant cables having thermosetting insulation and low emission of smoke and corrosive gases when affected by fire'.

BS 7846

5.4 Building complexity

The contribution of the cabling to the overall risk cannot be considered separately from that of the whole building (see paragraph 3.2 et seq).

See also Section 7 of this Guidance Note.

Section 6 — Safety Services

General 561
The requirements of Chapter 56 are applicable to supplies for fire pumps, booster pumps for sprinkler systems and risers, fire alarms, emergency lighting, and other equipment which may need to be used during a fire.
561-01-02

There may be a need to consult the distributor regarding switching arrangements of standby supplies, while Chapter 56 gives more detailed requirements concerning supplies for safety services.
313-02

Chap 56

Section 562 dealing with sources for safety services does not specifically mention vehicle starter batteries because these batteries are unlikely to fulfil the requirements prescribed and consequently should not be used.
562

Routing of wiring systems and the fire resistance (circuit integrity) of cables for safety services must receive careful attention.
563-01-01

Devices for protection against overload may be omitted for circuits which supply fire extinguishing devices. Protection against fault currents, however, must be provided. In such situations consideration shall be given to the provision of an overload alarm.
473-01-03

6.1 Emergency lighting and fire alarms 528
Design of these services is specifically mentioned in Regulation 110-01-01 as installations where the general requirements of BS 7671 are to be supplemented by the requirements or recommendations of other British Standards.
110-01-01

The relevant system design standards are:

    (i) BS 5266, Part 1 : 'Code of practice for the emergency lighting of premises other than cinemas and certain other specified premises used for entertainment'
BS 5266

(ii) BS 5266, Part 2 : 1998 'Code of practice for electrical low mounted systems for emergency use' BS 5266

(iii) BS 5266, Part 4 : 1999 'Emergency lighting code of practice for design, installation, maintenance and use of fibre systems' BS 5266
Note:
Part 3 deals with small power relays,
Part 5 deals with the specification for component parts of optical fibre systems,
Part 6 deals with non-electrical low mounted way guidance systems.

(iv) BS 5839-1 : 2002 'Fire detection and fire alarm systems for buildings, Part 1 : Code of practice for system design, installation, commissioning and maintenance' BS 5839-1
Note:
The 2002 edition takes into account changes in technology, custom and practice since the previous 1988 edition and introduces some significant changes.

(v) BS 5839, Part 6 : 'Code of practice for the design and installation of fire detection and alarm systems in dwellings' BS 5839 Part 6

(vi) BS 5839, Part 8 : 1998 'Fire detection and alarm systems for buildings. Code of practice for the design, installation and servicing of voice alarm systems'. BS 5839 Part 8

These standards have detailed recommendations regarding, for instance, power supplies and types of cable to be used. Reference to these standards is essential.

For the emergency lighting of cinemas and places of entertainment, reference should be made to the local licensing authority and to Code of Practice CP 1007 : 1955 'Maintained lighting for cinemas'. CP 1007

Wiring connecting a self-contained luminaire to the main supply is not considered to be an emergency lighting circuit. Mineral insulated cables and cables complying with categories of BS 6387 as required by the safety system standard are considered to be segregated e.g. category AWX or SWX fire alarm cables and category B emergency lighting cables.

| **6.2** | **Emergency lighting and fire alarms - Scotland** | In Scotland the requirements for emergency lighting and fire alarms are specified in the Scottish Office Technical Standards as follows: |

Emergency Lighting E7.7

The requirements of E7.7 will be met where emergency lighting is installed:

(i) in the case of cinemas, bingo halls, ballrooms, dance halls and bowling alleys, in accordance with CP 1007 : 1955

CP 1007

(ii) in the case of a single skin air supported structure falling within Class I to 4 in the table to BS 6661 : 1986, in accordance with BS 6661 : 1986 Section 5, Clauses 22, 24 and 25

BS 6661

Note:
Unfortunately BS 6661 has now been withdrawn and has not been replaced by another standard.

(iii) in the case of any other building, in accordance with BS 5266 Part 1 : 1988.

BS 5266

Automatic Fire Detection E10.1

The requirements of E10.1 will be met by an automatic fire detection and alarm system complying with BS 5839 Part 1 : 1988, Type L3. (See Note below paragraph 6.1, item(iv)).

BS 5839

An exception is made for dwellings where any storey is less than 200 m^2 in floor area. The dwelling may be provided with one or more mains operated smoke alarms with standby supplies, suitably located on each storey. Detailed requirements are given in the Technical Standards under Part E of 'Provisions deemed to satisfy the Standards'. See Appendix C.

6.3 Smoke detectors and alarms

The Smoke Detector Act 1991 requires that new dwellings are fitted with smoke detectors. Approved Document B, Fire Safety, 2000 Edition, issued under the Building Regulations 2000 Section 1 gives guidance on the installation of smoke detectors and alarms to facilitate escape from a building in case of fire. Section 1 of the Approved Document requires dwellings that are not protected by an automatic fire detection and alarm system in accordance with BS 5839 : Part 1 to at least L3 standard or to BS 5839 : Part 6 to at least system level of protection type LD3, grade F to be provided with smoke alarms in accordance with its particular requirements. These requirements are reproduced in Appendix B.

BS 5839

Note:
BS 5839, Part 1 : 1988 was replaced by BS 5839-1 : 2002 on the 15th July 2003. The categories (previously 'types') of system such as L1, L2 and L3 have been extended in number in the new code of practice to L1, L2, L3, L4 and L5. Persons involved in fire alarm systems will need to consult BS 5839-1 : 2002.

BS 5839-1

The equipment standard for smoke alarms and detectors is BS 5446-1 : 2000 'Fire detection and fire alarm devices for dwellings. Specification for smoke alarms'.

In Scotland, where the Building Standards (Scotland) Regulations 1990 apply, the requirements for smoke detectors in dwellings differ from the Building Regulations 2000 (see Section E4.1 of the Scottish Office Technical Standard). The two main differences are as follows:

(i) smoke detectors must be fitted to all dwellings in respect of new construction or change of use, where the application for building warrant was made after 23 July 1993

(ii) mains operated smoke alarms can be connected to the local lighting circuit or to a separate circuit. (This provides an indication of failure of supply to the alarm circuit.)

Part 6 of BS 5839 'Code of practice for the design and installation of fire detection and alarm systems in dwellings' gives recommendations for the planning, design and installation of fire detection and alarm systems in dwellings and dwelling units that are designed for the accommodation of a single family

BS 5839 Part 6

and in houses in multiple occupation. The recommendations are made both for new and existing dwellings.

6.4 Fireman's switches and other switches 476

BS 7671 is concerned with the provision of fireman's switches — Section 476 specifies where and how they are to be provided and Section 537 states the constructional requirements (see also Guidance Note 2: Isolation & Switching).

GN2

As there are also Statutory Regulations concerning fireman's switches the designer of the electrical installation should consult the relevant fire authority.

476-03-05
476-03-06
476-03-07
537-04-06

Other regulations in Chapter 46 and Section 476 also concern switching and the fire hazard, in particular in relation to electric cookers which should be controlled by a switch placed within easy reach.

However, the switch should be so positioned that in the event of a cooker fire the user need not lean over the cooker to operate the switch. The switch should therefore be installed to one side of the cooker but not more than 2 m from it and not in a cupboard where it would be inaccessible.

463-01-01
476-03-04

6.5 Static electricity

Measures to be taken to minimise the danger from static electricity are not included in BS 7671, but such measures are of particular importance in the handling of flammable liquids and dust. It is a highly specialised subject upon which guidance may be obtained from BS 5958 : 1991 'Code of practice for control of undesirable static electricity'. The Health Department's Health Guidance Note (HGN) 'Static discharges' deals with anti-static precautions in hospitals, and other information is available from the Health and Safety Executive.

BS 5958

6.6 Lightning 541

Chapter 11 specifically excludes lightning protection of buildings from the scope of BS 7671 but BS 7671 requires connection of the lightning protective system to the main earthing terminal of the installation. For information on protection against lightning, which includes fire among its possible effects, reference

110-02-01(ix)
413-02-02
541-01-03
542-04-01
BS 6651
BS 7430

should be made to BS 6651 'Code of practice for protection of structures against lightning' and BS 7430 'Code of practice for earthing'.

6.7 Exit signs

Approved Document B, Fire Safety, 2000 Edition

Approved Document B, Fire Safety, 2000 Edition issued under the Building Regulations 2000 Section 6 Paragraph 6.37 advises:

Except in dwellings, every escape route (other than those in ordinary use), should be distinctively and conspicuously marked by emergency exit sign(s) of adequate size complying with the Health and Safety (Safety Signs and Signals) Regulations 1996. In general, signs containing symbols or pictograms which conform to BS 5499 Part 1 'Fire safety signs, notices and graphic symbols. Specification for safety signs', satisfy these regulations. In some buildings additional signs may be needed to meet requirements under other legislation.

The Health and Safety (Safety Signs and Signals) Regulations 1996

The Health and Safety (Safety Signs and Signals) Regulations 1996 bring into force the Safety Signs Directive (92/58/EEC). This requires safety signs wherever there is a risk to health and safety that cannot be totally avoided or controlled by other means, and introduces the green running person/arrow signs.

The Health and Safety Executive advise that fire safety signs containing symbols or pictograms which conform to the requirements of BS 5499 will meet the requirements of the Health and Safety (Safety Signs and Signals) Regulations 1996 providing they continue to fulfil their purpose effectively.

BS 5499 also includes a combination sign, showing the format supplemented with the words exit and an arrow sign.

The Health and Safety Executive advise that all old-style text-only EXIT signs must be replaced by a pictogram or supplemented with new signs.

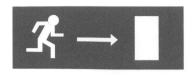 The Health and Safety
(Safety Signs and Signals)
Regulations format

 BS 5499-1 : 2002
format with supplementary
directional arrow signs

 BS 5499-1 : 2002
combination sign format

**6.8 Alarm
systems**

BS 7807 : 1995 'Code of practice for design,
installation and servicing of integrated systems
incorporating fire detection and alarm systems and/or
other security systems for buildings other than
dwellings', provides guidance on combined fire and
alarm systems.

Section 7 — Cable Selection

General

There are two basic factors to be considered when selecting cables with respect to fire hazard:

    (i) the contribution the cables themselves may make to the spread of the fire, evolution of smoke and fume emission

527-01-01

    (ii) the need for the integrity of safety circuits to be maintained under fire conditions.

131-05-02
563-01-02

Cable specifications are based on either construction or performance requirements.

7.1 Cable construction specifications

Cable construction requirements are contained in the following British standard specifications:

7.1.1 Mineral insulated

BS EN 60702-1 : 2002 'Mineral insulated cables and their terminations with a rated voltage not exceeding 750 V. Part 1 Cables'.

BS EN 60702-1

BS EN 60702-1 : 2002 applies to mineral insulated general wiring cables with copper or copper alloy sheath and copper conductors and with rated voltages of 500 V and 750 V. Provision is made for a corrosion resistant extruded outer covering over the copper sheath, when required. This outer covering is not specified for the purpose of electrical insulation of the metal sheath.

Requirements for terminations for use with these cables are specified in BS EN 60702-2.

The purpose of the standard is to specify mineral insulated cables that are safe and reliable when properly used, to state the manufacturing requirements and characteristics to achieve this, and to specify methods for checking conformity with those requirements.

The standard covers the voltage designations of 500 V cable (light duty grade) and 750 V cable (heavy duty grade).

The standard sets out requirements for the optional outer covering, which includes requirements for halogen-free covering and the thickness of the covering. The standard includes routine tests including a spark test on the outer covering. Sample tests are included such as flame retardance, emission of acidic and corrosive gases and smoke emission. Type tests such as fire resistance are included.

The current ratings for use when the cable is exposed to touch, in contact with combustible building materials, or when the cable is provided with an optional outer covering, give rise to sheath temperatures of 70 °C at an ambient temperature of 30 °C. Under controlled conditions and with the use of suitable terminations, when bare cable is not exposed to touch or in contact with combustible building materials the cable may be used continuously at temperatures up to 250 °C. BS 7671 provides current ratings for 105 °C operation, for other conditions the manufacturer should be consulted.

Table 53B

500 V grade cable includes the following conductor sizes:

— single and twin conductor cable up to 4.0 mm^2 csa
— three, four and seven conductor cable up to 2.5 mm^2 csa.

750 V grade provides for:

— single conductor cables up to 400 mm^2 csa
— two, three and four conductor cables up to 25 mm^2 csa
— seven conductor cables up to 4.0 mm^2 csa
— twelve conductor cables up to 2.5 mm^2 csa
— nineteen conductor cables up to 1.5 mm^2 csa.

7.1.2 Armoured thermosetting cables

BS 6724 : 1997 'Specification for 600/1000 V and 1900/3300 V armoured electric cables having thermosetting insulation and low emission of smoke and corrosive gases when affected by fire'.

BS 6724

The standard specifies requirements for 600/1000 V and 1900/3300 V armoured cables which produce lower levels of smoke and corrosive products under exposure to fire as compared with cables complying with BS 5467 and BS 6346.

The 600/1000 V cables included in the standard are as follows:

— single-core copper conductor
— single-core solid aluminium conductor
— two, three, four and five-core copper conductor
— two, three and four-core solid aluminium conductor
— multicore auxiliary copper conductor.

7.1.3 Non-armoured thermosetting cables

BS 7211 : 'Specification for thermosetting insulated cables (non-armoured) for electric power and lighting with low emission of smoke and corrosive gases when affected by fire'.

BS 7211

The standard specifies requirements for non-armoured cables with thermosetting insulation of rated voltage up to 450/750 V which, when assessed by tests specified in the standard, produce lower levels of smoke and corrosive gases under exposure to fire compared with pvc cables made to BS 6004.

Corrosive (and acid) gases are defined as those which are determined either as HCl or by measurement of pH and conductivity, as appropriate.

The standard specifies construction and test requirements, including tests relating to performance under fire conditions.

Types of cable included in the standard are:

— thermosetting insulated, non-sheathed, single-core cable 450/750 V with rigid copper conductors and stranded copper conductors

— thermosetting insulated, non-sheathed, single-core cable 300/500 V with rigid copper conductors and stranded copper conductors

— thermosetting insulated, single-core sheathed cable 450/750 V

— thermosetting insulated, twin, three-core, four-core and five-core circular sheathed cables 450/750 V

— thermosetting insulated, single-core, flat twin, and flat three-core sheathed cable with circuit protective conductor 300/500 V.

7.1.4 Cables with limited circuit integrity

BS 7629-1 : 1997 'Specification for 300/500 V fire resistant electric cables having low emission of smoke and corrosive gases when affected by fire. Multicore cables'.

BS 7629-1

BS 7629-2 : 1997 'Specification for 300/500 V fire resistant electric cables having low emission of smoke and corrosive gases when affected by fire. Multipair cables'.

BS 7629-2

These standards apply to cables with thermosetting insulation of rated voltage 300/500 V which conform to the performance requirements for cables required to maintain limited circuit integrity under those fire conditions of BS 6387 specified as B, W and X.

BS 6387

The cable standards specify constructional and performance requirements, and methods of test including tests relating to performance under fire conditions.

The cables are suitable for operation at a maximum sustained conductor temperature of 70 °C although the insulation is suitable for operation at higher temperatures. Use at a temperature not exceeding 90 °C is allowed for terminations within an enclosure providing the cable conductor temperature outside the enclosure does not exceed 70 °C.

The standards apply to cables which are 300/500 V, and
— two, three and four-core circular cables with uninsulated circuit protective conductor
— 7, 12 or 19 core with an uninsulated drain wire
— 1, 2, 5, 10 and 20 pairs having a collective metallic layer and drain wire.
They contain a metallic layer which provides electrostatic screening.

7.1.5 Armoured cables with limited circuit integrity

BS 7846 : 2000 'Electric cables 600/1000 V armoured fire resistant cables having thermosetting insulation and low emission of smoke and corrosive gases when affected by fire'.

The standard specifies requirements for construction and describes methods of test for armoured cables with thermosetting insulation of rated voltages 600/1000 V which, when assessed for fire resistance, meets the requirements of this standard. Cables included are intended for use in fixed installations in industrial areas, buildings and similar applications, where maintenance of power supply during a fire is essential and where the evolution of smoke and corrosive gases must be kept to a minimum.

For the purpose of the standard, cables are designated by category according to their special fire performance characteristics. The following categories are included:

Category F1 — resistance to fire alone

Category F2 — resistance to fire, resistance to fire with water, resistance to fire with mechanical shock, assessed separately

Category F3 — resistance to fire with mechanical shock and water assessed in combination

Corrosive (and acid) gases are defined as those which are determined as hydrochloric acid (HCl).

The cables are wire armoured and oversheathed as follows:

— two, three, four and five-core stranded copper conductor
— multicore auxiliary stranded copper conductor.

7.2 Cable performance specifications

There are British Standard performance specifications relating to cables for flammability, smoke density, emission of halogen acid gas and circuit integrity.

7.2.1 Flame spread

Flame spread is assessed either by BS EN 50265, for a single vertical cable, or by BS EN 50266 for vertical bunches of cables.

BS EN 50265-2-1 "Common test methods for cables under fire conditions. Test for resistance to vertical flame propagation for a single insulated conductor or cable. Procedures. 1 kW pre-mixed flame"

The test is carried out on a sample of cable 600 mm in length, suspended vertically between two clamps 550 mm apart in a draught-free enclosure and exposed to a heat source in the form of a 1 kW pre-mixed gas flame. The burner is arranged at an angle of 45° such that it impinges on the cable 475 mm below the top clamp. The requirement is that after a specified time, which increases as the cable diameter increases, the burner is removed and after all burning has ceased the sample is wiped clean and the charred or affected portion of the cable shall not have reached within 50 mm of the lower edge of the top clamp or more than 540 mm downwards from the same lower edge.

BS EN 50265-2-2 "Common test methods for cables under fire conditions. Test for resistance to vertical flame propagation for a single insulated conductor or cable. Procedures. Diffusion flame"

This part is similar in test method to Part 2-1 except that a specified burner of lower output energy is used. The method specified in Part 2-1 is unsuitable for wires/cables with solid conductors less than 0.8 mm diameter or stranded conductors 0.1 to 0.5 mm² because it is likely that such small conductors will melt.

BS EN 50266 "Common test methods for cables under fire conditions. Test for vertical flame spread of vertically-mounted bunched wires or cables."

BS 50266 is in 6 parts. Part 1 describes the test apparatus. Parts 2-1 to 2-5 describe different categories of test procedure, all of which assess the ability of layers and bunches of cables to restrain flame propagation in defined conditions regardless of their application, i.e. power, telecommunications, etc.

The categories are defined and distinguished by test duration, use or one or two burners, the way the cables are mounted and the volume of non-metallic material (NMV) of the samples under test; they are not necessarily related to different safety levels in actual cable installations. Below is an indicative summary table of the differences.

Part of BS EN 50266	Category	NMV (litres per metre of cable)	No of burners	Cable conductor size (mm²)	Ladder mounting	Test duration (minutes)
2-1	A F/R	7	1	35 +	Front and rear	40
2-2	A	7	1 or 2	All	Front. Wider ladder for larger cables	40
2-3	B	3.5	1	All	Front	40
2-4	C	1.5	1	All	Front	20
2-5	D	0.5	1	Up to 12	Front	20

The ignition source is one or two ribbon type propane/air gas burners with a fuel input of 73.7 MJ/h (70000 BTU/h). The burner is arranged horizontally at the foot of the ladder and the flame applied for the times shown. Cables burning to 2.5 m above the burner are classified as failed.

There are a number of significant features affecting propagation listed in the standard namely, ratio of combustible to non-combustible material, the volume of combustible gases evolved and the temperature at which they ignite, volume of air passing through the installation, and cable construction e.g. armoured or unarmoured.

It is noted that the configuration of an actual cable installation will affect the level of flame spread occurring in a actual fire.

7.2.2 Measurement of smoke density

BS EN 50268 : 2000 'Common test methods for cables under fire conditions. Measurement of smoke density of cables burning under defined conditions'.

BS EN 50268 specifies a method of test for measurement of smoke density from cables burning under defined conditions. It is suitable for electric insulated conductor or cable, or optical cables. Part 1 details the apparatus.

Part 2 provides details of the test procedure to be employed for the measurement of the smoke density of the products of combustion of electric cables burning under defined conditions, using the test apparatus given in Part 1.

Testing is based on a number of burning cable samples, 1 m in length, mounted horizontally over a bath containing 1 litre +/- 0.01litre of alcohol (90% +/- 1% ethanol, 4% +/- 1% methanol, 6% +/- 1% water) as the fire source. The density of the smoke generated is monitored by a light beam and photocell arranged at a height of 2.15 m horizontally across the enclosure which is a 3.0 m sided cube. A fan, suitably screened, is provided to prevent stratification of the smoke.

A method of sample selection and assembly is specified and guidance is given on the acceptable level of smoke density, expressed as transmittance levels.

7.2.3 Measurement of halogen acid gas emission

BS EN 50267-2-1 : 1999 'Common test methods for cables under fire conditions. Tests on gases evolved during combustion of materials from cables. Procedures. Determination of the amount of halogen acid gas.'

This British Standard using apparatus in BS EN 50267-1 describes a method for the determination of the amount of halogen acid gas, evolved during the combustion of compounds based on halogenated polymers and compounds containing halogenated additives taken from cable constructions i.e. the construction of the cable.

Determination of the amount of halogen acids, especially hydrogen chloride (HCl), is carried out by heating a sample of material in a stream of dry air and absorbing the gases in 0.1 M sodium hydroxide solution. The amount of halogen acid is then determined analytically and expressed in percentage of hydrogen chloride.

An alternative method, using determination of pH and conductivity is described in BS EN 50267-2-2, and is used for some harmonised cables.

7.3 Circuit integrity

BS 6387 : 1994 'Specification for performance requirements for cables required to maintain integrity under fire conditions'.

The standard specifies performance requirements and gives test methods for mechanical and fire tests applicable to cables rated at voltages not exceeding 450/750 V and for mineral insulated cables conforming to BS 6207. (BS EN 60702-1 : 2002)

The cables are intended to be used for wiring and interconnection where it is required to maintain circuit integrity under fire conditions for longer periods than can be achieved with cables of conventional construction.

Those requirements of cables, related to characteristics required to enable circuit integrity to be maintained under fire conditions are specified.

Cables tested to this standard are categorised under three separate test conditions.

The first letter indicates a resistance to fire alone:

A   650 °C for 3 hours
B   750 °C for 3 hours
C   950 °C for 3 hours
S   950 °C for 20 minutes.

The second letter **W** indicates resistance to fire at 650 °C with water.

The third letter indicates resistance to fire with mechanical shock. Mechanical shocks are applied to the cable at specified temperatures of:

X   650 °C
Y   750 °C
Z   950 °C.

The fire related properties required by the cable standards are summarised in Table 7.3.

TABLE 7.3

Cable Standard and type		Fire related properties	
BS EN 60702	Mineral insulated cables with a rated voltage not exceeding 750 V	BS EN 50265 and -2-1	Tests on electric cables under fire conditions
		BS EN 50268-2 (for cables with zero-halogen coverings)	Measurement of smoke density of electric cables burning under defined conditions
		BS EN 50267 and -2-1 (for cables with zero-halogen coverings)	Gases evolved during combustion of electric cables
		BS 6387 Cat C, W & Z	Performance requirements for cables required to maintain integrity under fire conditions
BS 6724	Armoured cables for electricity supply having thermosetting insulation with low emission of smoke and corrosive gases when affected by fire	BS EN 50265-1 and -2-1	Tests on electric cables under fire conditions - single cable
		BS EN 50266-2-4	Tests on electric cables under fire conditions - bunched cables
		BS EN 50268-2	Measurement of smoke density of electric cables burning under defined conditions
		BS EN 50267-1 and -2-1	Gases evolved during combustion of electric cables
BS 7211	Thermosetting insulated cables (non-armoured) for electric power and lighting with low emission of smoke and corrosive gases when affected by fire	BS EN 50265-1 and -2-1	Tests on electric cables under fire conditions
		BS EN 50268-2	Measurement of smoke density of electric cables burning under defined conditions
		BS EN 50267-1 and -2-1	Gases evolved during combustion of electric cables

Cable Standard and type		Fire related properties	
BS 7629	Thermosetting insulated cables with limited circuit integrity when affected by fire	BS EN 50265-2-1	Tests on electric cables under fire conditions
		BS EN 50268-2	Measurement of smoke density of electric cables burning under defined conditions
		BS 6387 Cat B, W & X	Performance requirements for cables required to maintain integrity under fire conditions
		BS EN 50267-2-1	Gases evolved during combustion of electric cables
BS 7846	600/1000 V armoured electric cables having low emissions of smoke and corrosive gases when affected by fire	BS EN 50265-2-1	Tests on electric cables under fire conditions - single cable
		BS EN 50266-2-4	Tests on electric cables under fire conditions - bunched cables
		BS EN 50268-2	Measurement of smoke density of electric cables burning under defined conditions
		BS EN 50267-2-1	Gases evolved during combustion of electric cables
		BS 7846 Cat F1, F2 or F3C	Performance requirements for cables required to maintain integrity under fire conditions

7.4	**System requirements**	The following system standards include cabling requirements:	

	BS 5266	Emergency lighting	BS 5266
	BS 5839	Fire detection and alarm systems for buildings	BS 5839
	BS 6266	Code of practice for fire protection for electronic data processing installations.	BS 6266

These requirements are briefly summarised in Table 7.4.

7.4.1 Emergency lighting cables

Fire withstand of cables

BS 5266 : Part 1 advises that wiring connecting a self-contained luminaire to the normal supply is not considered to be an emergency lighting circuit. Wiring of escape lighting to the standby power supply should possess inherently high resistance to attack by fire and to physical damage, or have additional fire protection. Cables considered to have inherently high resistance to fire are:

Mineral insulated copper sheathed cable (micc) to BS EN 60702, cables to BS 7629 and BS 7846, do not require segregation.

Additional protection against fire can be provided by burying cables in the building structure or situating them where there is a negligible fire risk and separating from a significant fire risk by a suitable wall partition or floor.

Segregation

Escape lighting circuits shall be completely segregated from all other cables in accordance with BS 5266.

Escape lighting is that part of the emergency lighting provided to illuminate the escape route.

Escape lighting system cables should either be separated from the cables of other services by a minimum distance of 300 mm between centre lines of the cables, or be mineral insulated copper-sheathed cable, with or without oversheath, in accordance with BS EN 60702-1, rated for exposure to touch. (For example, see Table 4J1A in BS 7671, which gives the current-carrying capacities of mineral insulated cables bare and exposed to touch when the tabulated values are multiplied by 0.9 in order to limit the sheath to a safe operating temperature.) Also acceptable is cable

BS 5266

BS 5266 : Part 1

BS EN 60702 Table 4J1A

complying with BS 6387 and assessed as suitable for use where separation is not provided under the 'BASEC Certification of assessment' scheme.

The escape lighting system cable(s) should be completely enclosed when the cover of any associated trunking or channel is in place.

Ducting, trunking or channel reserved for escape lighting system cable should be marked to indicate this reservation.

Multicore cables should not be used to service both escape lighting and any other circuit.

7.4.2 Fire detection and alarm systems

Fire withstand of cables

The requirements for the fire withstand of cables have been completely revised in the new BS 5839-1: 2002. All persons involved in fire alarm systems will need to consult this standard to satisfy themselves of compliance. The information given here is only intended as a brief guide.

Two of the principal changes made within the revised standard are:

- the use of fire resisting cables is now recommended for all manual call point and automatic fire detector circuits. The use of fire resisting cables is also recommended for all mains power supply circuits;
- two different levels of resistance of cables to damage during a fire are recognised, and recommendations for application of each type are provided.

The standard includes a commentary, extracts of which are given here:

This standard makes recommendations for two levels of fire resistance of fire resisting cable systems, termed "standard" and "enhanced", according to the type of building and fire alarm system installed:

- the use of cables with "standard" fire resistance is recommended for general use;
- the use of cables with "enhanced" fire resistance is recommended for systems, in particular building types, in which cables might need to operate correctly during a fire for periods in excess of those normally required for single phase evacuation of a building. Examples are unsprinklered high-rise

buildings with phased evacuation arrangements and premises of such a nature or size that areas remote from the fire could continue to be occupied for a prolonged duration during a fire that might then damage cables serving parts of the fire alarm system in occupied areas.

Cables capable of complying with the recommendations for "standard" fire resistance are expected to include some that have been commonly used for many years for circuits in fire alarm systems that must operate for a prolonged period during a fire, in accordance with BS 5839-1: 1988, without any evidence from real fires that satisfaction of the objectives of a fire alarm system necessitates a higher performance. It is recognized, however, in the revision of BS 5839-1, that the level of fire resistance described as "enhanced" is desirable in certain systems in particular building types, although unnecessary for most systems. Cables complying with the recommendations for "enhanced" fire resistance are expected to include, amongst other types, some mineral insulated copper sheathed cables conforming to BS EN 60702-1. *Note by the IEE. The two test methods (BS 8434-1 and BS 8434-2) have been published to satisfy the requirements relating to the two classes of cable. It is strongly advised to consult with the manufacturers concerning applicability.*

BS EN 60702

BS 8434

The circuits of fire alarm systems must be segregated from the cables of other circuits to minimize any potential for other circuits to cause malfunction of the fire alarm system arising from:

528-01-04

- breakdown of cable insulation of other circuits and/or fire alarm circuits;
- a fire caused by a fault on another circuit;
- electromagnetic interference to any fire alarm circuit as a result of the proximity of another circuit;
- damage resulting from the need for other circuits to be installed in, or removed from, ducts or trunking containing a fire alarm circuit.

In order to facilitate identification of fire alarm circuits, cables should preferably be red in colour, unless another form of colour coding is appropriate. By this means, the possible need for appropriate segregation can be identified, and there will be less likelihood of inadvertent manual interference with the circuits of fire alarms(eg during work on other electrical circuits).

Refer to BS 5839-1: 2002 for the detailed recommendations.

7.4.3 Electronic data processing   BS 6266 provides general guidance on the selection of cables. In critical situations cables are required to meet BS 6387 category AWZ, or mineral insulated cables to BS EN 60702 or cables to BS 7629 should be used.

BS 6266

BS 6387
BS EN 60702
BS 7629

The requirements of these standards are summarised in Table 7.4.

TABLE 7.4

Standard		Example Requirement
BS 5266 Part 1	Emergency Lighting Code of practice for the emergency lighting of premises other than cinemas and certain other specified premises used for entertainment	In general, the following cables must be used for connecting luminaires to the power supply, 8.2.2(a) those with inherently high resistance to attack by fire: (a) micc cable to BS EN 60702 (b) cables to BS 6387 category B e.g. BS 7629 and BS 7846 Cables as follows require additional protection 8.2.2(b): (c) pvc cables to BS 6004 in steel conduit (d) pvc cables to BS 6004 in rigid pvc conduit, classification 405/100,000 or 425/100,000 to BS 6099 Section 2.2 (e) pvc or XLPE insulated and sheathed steel wire armoured cables to BS 6346 or BS 5467.
BS 5839 Part 1	Fire detection and alarm systems for buildings. Code of practice for system design, installation, commissioning and maintenance.	Refer to paragraph 7.4.2 within this Guidance Note.
BS 6266	Fire protection for electronic data processing installations	Cables to comply with BS EN 50265, e.g. all pvc cables. Where appropriate, power cables should comply with BS 6724 (armoured thermosetting low smoke and gas) 4.2.1.5. In critical installations power cables to meet category AWZ of BS 6387 (note to 4.2.3.5) e.g. micc to BS 6207 (BS EN 60702-1) and cables to BS 7629.

Section 8 — Maintenance

General

Regulation 4(3) of the Electricity at Work Regulations 1989 requires that:

As may be necessary to prevent danger, all systems shall be maintained so as to prevent, so far as is reasonably practicable, such danger.

Electrical installations that are not regularly maintained may present a particular hazard with respect to fire.

BS 7671 requires that the design and construction of the installation shall provide for accessibility to equipment for inspection, testing, maintenance and repair.

131-12-01

An assessment should be made of the likely frequency and quality of maintenance that the installation might reasonably receive, and the design of the installation must take this into account.

341-01-01

Similarly, an installation itself must be maintained as intended by the designer and installer. The details of necessary or presumed maintenance must be passed on to the user, as required by Section 6 of the Health and Safety at Work etc Act 1974. Guidance is given on the preparation of technical manuals in BS 4884 and BS 4940.

BS 4884
BS 4940

In order to comply with the Electricity at Work Regulations (Regulation 4(3)), and for the avoidance of danger generally, periodic inspection and testing of an electrical installation is necessary, and BS 7671 includes requirements for this. It should be carried out at an appropriate frequency, as intended by the designer, and as endorsed upon the Electrical Installation Certificate.

731-01-01
732-01-01
Appx 6

One of the consequences of insufficient inspection, testing and maintenance of an installation can be overheating leading to fire.

Further guidance on the inspection of electrical installations is given in Guidance Note 3 : Inspection & Testing. Discussed below are aspects of inspection related to fire and the electrical installation.

GN3

8.1 Overheating

Accessories, switchgear, etc, and particularly plastic accessories, should be scrutinised for signs of overheating. This is often an indication of a loose connection, for example in a socket-outlet or in the plug. It is essential that any accessory, including socket-outlets, that shows any sign of scorching is replaced.

External influences and any changes thereof must be considered during the inspection, particularly with respect to the suitability of accessories or equipment for the environment, including any environmental factors which may not have been apparent to the designer — contamination of creepage surfaces of equipment must be borne in mind if accessories of an inadequate IP rating are fitted in adverse environments (wet, damp), e.g. in air handling ducts. These aspects are of obvious importance where there has been a change of use of the premises.

8.2 Structural changes

Changes to the building can adversely affect all equipment, but especially cabling. Covering cables with thermal insulation will reduce their current-carrying capacity, and certain cables such as bare sheathed mineral insulated should not be in contact with flammable materials. This can equally apply to bare earthing conductors, which may be designed to operate during fault clearance at high temperatures. In these circumstances, particular care has to be taken to ensure that no flammable materials are in contact with the cables or conductors.

8.3 Extension leads and adaptors

The use of extension leads and adaptors must be discouraged. It is important that equipment and appliances are only used with a lead allowed by the particular product standard, and this should be plugged directly into a socket-outlet complying with the requirements of BS 7671. The relatively high loop impedance of extension leads can result in incorrect operation of overcurrent devices, leading to overheating of the extension lead. Incorrectly sized extension leads are an obvious hazard, and coiling of extension leads reduces their current rating. Guidance on extension leads is given in the 'Code of Practice for In-Service Inspection and Testing of Electrical Equipment' published by the IEE.

There may be further problems if extension leads are used in conjunction with office furniture and IT equipment. See paragraph 8.4 below.

8.4 IT equipment

The requirements of Section 607 of BS 7671 need to be consulted.

Regulation 607-03-01 states:- 607-03-01
'For a final circuit with a number of socket-outlets or connection units intended to supply several items of equipment, where it is known or reasonably to be expected that the total protective conductor current in normal service will exceed 10 mA, the circuit shall be provided with a high integrity protective conductor connection complying with the requirements of Regulations 607-02 and 607-04.'

The regulation lists arrangements of final circuits which are acceptable. These include:

(i) a ring final circuit with a ring protective conductor.

(ii) a radial final circuit with a single protective conductor:

  (a) the protective conductor being connected as a ring; or

  (b) a separate protective conductor being provided at the final socket-outlet by connection to the metal conduit or ducting; or

  (c) where two or more similar radial circuits supply socket-outlets in adjacent areas and are fed from the same distribution board, and they also have identical means of short-circuit and overcurrent protection and circuit protective conductors of the same cross-sectional area, then a second protective conductor may be provided at the final socket-outlet on one circuit by connection to the protective conductor of the adjacent circuit.

This is obviously relevant for extension leads where the sum of the protective conductor currents may well exceed 10 mA, and in the event of a loss of connection with earth in the extension lead there will be a considerable shock risk at each piece of equipment on the end of the extension lead.

Installations which subsequent to the initial design and construction have been fitted with considerable amounts of IT equipment, will need to be inspected to determine if they comply with Section 607.

8.5 Harmonics

The extensive use of switched mode power supplies can result in overloading of neutral conductors because of harmonic distortion of the load currents. Third harmonics and multiples drawn by such devices do not cancel out in three-phase supplies but summate. Neutral currents and inrush currents need reassessing after substantial change of use.

524-02-02

Appendix A: Safety Service and Cable Standards

System specifications

BS 5839 :   Fire detection and alarm systems for buildings
-1 : 2002 Code of practice for system design,
commissioning and maintenance
Part 6: 1995 Code of practice for the design
and installation of fire detection and alarm
systems in dwellings
Part 8: 1998 Fire detection and alarm systems
for buildings. Code of practice for the design,
installation and servicing of voice alarm
systems.

Note:

BS 5839-1 : 2002 superseded BS 5839 Part 1 : 1988 on 15 July
2003. Projects designed to BS 5839 Part 1 : 1988 may, by
agreement between contractual parties, be completed to that
edition.

This new edition of BS 5839-1 takes into account changes in
technology, custom and practice since the publication of the
1988 edition and introduces some significant changes.

The principal changes made within the revision are as follows:

a) the term "category" (of system) is now used instead of
"type" in the description of system objective/area of
coverage (e.g. a full property protection system was
previously described as type P1, but is now described as
Category P1); this is to distinguish this term from the less
precise use of the term "type" to describe the principles of
operation of the system (e.g. conventional, addressable,
heat or smoke, etc,);

b) the Categories (previously "types") of system defined in
this code of practice have been extended in number to give
recognition to systems that are designed to satisfy specific
fire safety objectives or enhance life safety, but that cannot
be classified within the system types previously defined;

c) the need for the level of protection to be based on a fire
risk assessment by a competent person is recognized;

d) the use of automatic fire detection as part of a fire
engineering solution is recognized;

e) the use of new technologies, such as multi-sensor detection, is addressed;

f) new methods of fire detection, including carbon monoxide detection and video smoke detection, are taken into account;

g) the maximum distance of travel to a manual call point has been increased;

h) greater flexibility is recommended in respect of minimum sound pressure levels;

i) a new section, devoted to the limitation of false alarms by appropriate system design, system management and improved technology has been added;

j) the distinction between different categories of false alarms, particularly those that result from environmental influences or fire-like phenomena, and those that result from equipment faults, is recognized;

k) two different levels of resistance of cables to damage during the course of a fire are recognized, and recommendations for application of each type are provided;

l) the use of fire resisting cables is now recommended for all manual call point and automatic fire detector circuits. The use of fire resisting cables is also recommended for all mains power supply circuits;

m) recommendations for networked systems, particularly in respect of cable types, are included;

n) recommendations for fire warning systems for people with impaired hearing have been included;

o) the code of practice has been simplified by the use of practice specification format, in which commentary on relevant principles is followed by short, succinct recommendations, and by new diagrams. This is intended to make the code of practice less ambiguous, simpler for the non-specialist to apply and compliance of installations more straightforward to audit;

p) recommendations for servicing and maintenance of systems, including the periods at which routine servicing should be carried out, have been revised;

q) a new informative annex has been added to give advice on the Categories of system typically installed in a variety of occupancies;

r) separate recommendations are no longer provided for control and indicating equipment and power supplies for small manual systems; such systems should comply with all recommendations of the code of practice;

s) the term "deviation" has been replaced with the term "variation", to avoid any negative connotation associated with the term used to describe an aspect of system design that, for sound reasons, does not comply with the recommendations of this standard;

t) the single certificate of commissioning has been replaced with separate certificates for design, installation, commissioning, acceptance and verification.

BS 5266 :	Emergency lighting Part 1 : 1988 Code of practice for the emergency lighting of premises other than cinemas and certain other specialised premises used for entertainment. Part 2 : 1998 Code of practice for electrical low mounted systems for emergency use. Part 4 : 1999 Emergency lighting code of practice for design, installation, maintenance and use of fibre systems.

Note:
Part 3 deals with small power relays,
Part 5 deals with the specification for component parts of optical fibre systems,
Part 6 deals with non electrical low mounted way guidance systems.

BS 6266 : 1992	Code of practice for fire protection for electronic data processing installations.
BS 7807 : 1995	Code of practice for design, installation and servicing of integrated systems incorporating fire detection and alarm systems and/or other security systems for buildings other than dwellings.

Cable performance specifications

BS EN 50265-1 : 1999	Common test methods for cables under fire conditions. Test for resistance to vertical flame propagation for a single insulated conductor or cable. Apparatus.
BS EN 50265-2-1 : 1999	Common test methods for cables under fire conditions. Test for resistance to vertical flame propagation for a single insulated conductor or cable. Procedures. 1 kW pre-mixed flame.
BS EN 50265-2-2 : 1999	Common test methods for cables under fire conditions. Test for resistance to vertical flame propagation for a single insulated conductor or cable. Procedures. Diffusion flame.
BS EN 50266-1 : 2001	Common test methods for cables under fire conditions. Test for vertical flame spread of vertically mounted bunched wires or cables. Apparatus.
BS EN 50266-2-1 : 2001	Common test methods for cables under fire conditions. Test for vertical flame spread of vertically mounted bunched wires or cables. Procedures. Category A F/R.
BS EN 50266-2-2 : 2001	Procedures. Category A.
BS EN 50266-2-3 : 2001	Procedures. Category B.
BS EN 50266-2-4 : 2001	Procedures. Category C.
BS EN 50266-2-5 : 2001	Procedures. Category D.
BS EN 50267-1 : 1999	Common test methods for cables under fire conditions. Tests on gases evolved during combustion of materials from cables. Apparatus.
BS EN 50267-2-1 : 1999	Common test methods for cables under fire conditions. Tests on gases evolved during combustion of materials from cables. Procedures. Determination of the amount of halogen acid gas.
BS EN 50267-2-2 : 1999	Procedures. Determination of degree of acidity of gases for materials by measuring pH and conductivity.

BS EN 50268-1 : 2000	Common test methods for cables under fire conditions. Measurement of smoke density of cables burning under defined conditions. Apparatus.
BS EN 50268-2 : 2000	Common test methods for cables under fire conditions. Measurement of smoke density of cables burning under defined conditions. Procedure.
BS 8434-1 : 2003	Methods of test for assessment of the fire integrity of electric cables - Test for unprotected small cables for use in emergency circuits —BS EN 50200 with addition of water spray
BS 8434-2 : 2003	Methods of test for assessment of the fire integrity of electric cables — Test for unprotected small cables for use in emergency circuits — BS EN 50200 with a 930 °C flame and with water spray

Cable specifications

Mineral insulated cables

BS EN 60702-1 : 2002, IEC 60702-1 : 2002	Mineral insulated cables and their terminations with a rated voltage not exceeding 750 V. Cables.

Thermosetting cables

BS 5467 : 1997	Specification for 600/1000 V and 1900/3300 V armoured electric cables having thermosetting insulation.
BS 6724 : 1997	Specification for 600/1000 V and 1900/3300 V armoured electric cables having thermosetting insulation with low emission of smoke and corrosive gases when affected by fire.
BS 7211 :1998	Specification for thermosetting insulated cables (non-armoured) for electric power and lighting with low emission of smoke and corrosive gases when affected by fire.

BS 7629-1 : 1997	Specification for 300/500 V fire resistant electric cables having low emission of smoke and corrosive gases when affected by fire. Multicore cables.
BS 7629-2 : 1997	Specification for 300/500 V fire resistant electric cables having low emission of smoke and corrosive gases when affected by fire. Multipair cables.
BS 7846 : 2000	Electric cables 600/1000 V armoured fire resistant cables having low emission of smoke and corrosive gases when affected by fire.

PVC cables

BS 6004 : 2000	Electric cables. PVC insulated (non-armoured) cables for voltages up to and including 450/750 V, for electric power, lighting and internal wiring.
BS 6231 : 1998	Specification for pvc insulated cables for switchgear and controlgear wiring.
BS 6346	Specification for 600/1000 V and 1900/3300 V armoured electric cables having pvc insulation.

Cable selection

BS 7540 : 1994	Guide to the use of cables with a rated voltage not exceeding 450/750 V .

British Standards are available from:

BSI Customer Services
389 Chiswick High Road
London   W4 4AL

Tel:	General Switchboard:	020 8996 9000
	For ordering:	020 8996 7000
	For information or advice:	020 8996 7111
	For membership:	020 8996 7002
Fax:	For ordering:	020 8996 7001
	For information or advice:	020 8996 7048

Appendix B: Section 1 of Approved Document B, 2000 edition, Fire Safety, Guidance on the Building Regulations
(For Scotland see Appendix C.)

AUTOMATIC SMOKE DETECTION AND ALARMS

1.2   In most houses the installation of smoke alarms or automatic fire detection and alarm systems, can significantly increase the level of safety by automatically giving an early warning of fire.

1.3   If houses are not protected by an automatic fire detection and alarm system in accordance with the relevant recommendations of BS 5839 : Part 1 to at least L3 standard, or BS 5839 : Part 6 to at least Grade E type LD3 standard, they should be provided with a suitable number of smoke alarms installed in accordance with the guidance in paragraphs 1.4 to 1.22 below.

Note:
BS 5839, Part 1 : 1988 was replaced by BS 5839-1 : 2002 on the 15th July 2003. The categories (previously 'types') of system such as L1, L2 and L3 have been extended in number in the new code of practice to L1, L2, L3, L4 and L5. Persons involved in fire alarm systems will need to consult BS 5839-1 : 2002.

BS 5839-1

1.4   The smoke alarms should be mains-operated and conform to BS 5446 : Part 1. They may have a secondary power supply such as a battery (either rechargeable or replaceable) or capacitor. More information on power supplies is given in clause 12 of BS 5839 : Part 6.

Note:
BS 5446 : Part 1 covers smoke alarms based on ionisation chamber smoke detectors and optical or photo-electric smoke detectors. The different types of detector respond differently to smouldering and fast flaming fires. Either type of detector is

generally suitable. However, the choice of detector type should, if possible, take into account the type of fire that might be expected and the need to avoid false alarms.

BS 5839 : Part 6 suggests that, in general, optical smoke alarms should be installed in circulation spaces such as hallways and landings, and ionisation chamber detectors may be the more appropriate type in rooms, such as the living room or dining room where a fast burning fire may present a greater danger to occupants than a smouldering fire.

Large dwelling houses

1.5   A house may be regarded as large if any of its storeys exceed 200 m².

1.6   A large house of more than 3 storeys (including basement storeys) should be fitted with an L2 system as described in BS 5839 : Part 1, 1988 except that the provisions in clause 16.5 regarding duration of the standby supply need not be followed.

> **Note:**
> BS 5839, Part 1 : 1988 was replaced by BS 5839-1 : 2002 on the 15th July 2003. The categories (previously 'types') of system such as L1, L2 and L3 have been extended in number in the new code of practice to L1, L2, L3, L4 and L5. Persons involved in fire alarm systems will need to consult BS 5839-1 : 2002.

BS 5839-1

However with unsupervised systems, the standby supply should be capable of automatically maintaining the system in normal operation (though with audible and visible indication of failure of the mains) for 72 hours, at the end of which sufficient capacity remains to supply the maximum alarm load for at least 15 minutes.

1.7   A large house of no more than 3 storeys (including basement storeys) may be fitted with an automatic fire detection and alarm system of Grade B type LD3 as described in BS 5839 : Part 6, instead of an L2 system.

Loft conversions

1.8   Where a loft in a one or two storey house is converted into habitable accommodation following the guidance in paragraphs 1.37 to 1.46 below, rather than by forming a fully protected internal stairway, then an automatic smoke detection and alarm system based on linked smoke alarms should be installed (see paragraph 2.26).

Automatic smoke detection and alarms

1.9   2.26 smoke alarms should be fitted as described in Section 1.

Sheltered housing

1.10   The detection equipment in a sheltered housing scheme with a warden or supervisor should have a connection to a central monitoring point (or central alarm relay station), so that the person in charge is aware that a fire has been detected in one of the dwellings and can identify the dwelling concerned. These provisions are not intended to be applied to the common parts of a sheltered housing development, such as communal lounges, or to sheltered accommodation in the Institutional or other residential purpose groups.

Installations based on smoke alarms

1.11   Smoke alarms should normally be positioned in the circulation spaces, sleeping spaces and places where fires are most likely to start (e.g. kitchens and living rooms) to pick up smoke in the early stages, while also being close enough to bedroom doors for the alarm to be effective when occupants are asleep.

1.12   In a house (including bungalow) there should be at least one smoke alarm on every storey.

1.13   Where more than one smoke alarm is installed they should be linked so that the detection of smoke by one unit operates the alarm signal in all of them. The manufacturer's instructions about the maximum number of units that can be linked should be observed.

1.14   Smoke alarms should be sited so that:

a   there is a smoke alarm in the circulation space within 7.5 m of the door to every habitable room

b   where the kitchen area is not separated from the stairway or circulation space by a door, there is an interlinked heat detector in the kitchen, in addition to whatever smoke alarms are needed in the circulation space(s) c they are ceiling mounted and at least 300 mm from walls and luminaires (unless, in the case of luminaires, there is test evidence to prove that the proximity of the luminaires will not adversely affect the efficiency of the detector). Units designed for wall mounting may also be used, provided that the units are above the level of doorways opening into the space and they are fixed in accordance with the manufacturer's instructions, and

d the sensor in ceiling mounted devices is between 25 mm and 600 mm below the ceiling (25-150 mm in the case of heat detectors).

NOTE: This guidance applies to ceilings that are predominantly flat and horizontal.

1.15   It should be possible to reach the smoke alarms to carry out routine maintenance, such as testing and cleaning, easily and safely. For this reason smoke alarms should not be fixed over a stair shaft or any other opening between floors.

1.16   Smoke alarms should not be fixed next to or directly above heaters or air conditioning outlets. They should not be fixed in bathrooms, showers, cooking areas or garages, or any other place where steam, condensation or fumes could give false alarms.

Smoke alarms should not be fitted in places that get very hot (such as a boiler room), or very cold (such as an unheated porch). They should not be fixed to surfaces which are normally much warmer or colder than the rest of the space, because the temperature difference might create air currents which move smoke away from the unit.

A requirement for maintenance cannot be made as a condition of passing plans by the building control body. However, the attention of developers and builders is drawn to the desirability of providing the occupants with information on the use of the equipment, and on its maintenance (or guidance on suitable maintenance contractors).

Note that BS 5839 : Part 1 and Part 6 recommend that the users should receive the operation and maintenance instructions of the alarm system.

Power supplies

1.17   The power supply for a smoke alarm system should be derived from the dwelling's mains electricity supply. The mains supply to the smoke alarm(s) should comprise a single independent circuit at the dwelling's main distribution board (consumer unit). If the smoke alarm installation does not include a standby power supply, no other electrical equipment should be connected to this circuit (apart from a dedicated monitoring device installed to indicate failure of the mains supply to the smoke alarms - see below).

1.18   A smoke alarm, or smoke alarm system, that includes a standby power supply or supplies, can operate during mains failure. It can therefore be connected to a regularly used local lighting circuit. This has the advantage that the circuit is unlikely to be disconnected for any prolonged period.

1.19   Devices for monitoring the mains supply to the smoke alarm system may comprise audible or visible signals on each unit or on a dedicated mains monitor connected to the smoke alarm circuit. The circuit design of any mains failure monitor should avoid any significant reduction in the reliability of the supply, and should be sited so that the warning of failure is readily apparent to the occupants. If a continuous audible warning is given, it should be possible to silence it.

1.20   The smoke alarm circuit should preferably not be protected by any residual current device (RCD). However, if electrical safety requires the use of a RCD, then either:

a   the smoke alarm circuit should be protected by a single RCD which serves no other circuit, or

b   the RCD protection of a smoke alarm circuit should operate independently of any RCD protection for circuits supplying socket-outlets or portable equipment.

1.21   Any cable suitable for domestic wiring may be used for the power supply and interconnection to smoke alarm systems. It does not need any particular fire survival properties. Any conductors used for interconnecting alarms (signalling) should be readily distinguishable from those supplying mains power, e.g. by colour coding.

Note: Smoke alarms may be interconnected using radio-links, provided that this does not reduce the lifetime or duration of any standby power supply.

1.22   Other effective, though possibly more expensive, options exist. For example, the mains supply may be reduced to extra-low voltage in a control unit incorporating a standby trickle-charged battery, before being distributed at that voltage to the alarms.

WALL AND CEILING LININGS

Classification of linings

7.1   Subject to the variations and specific provisions described in paragraphs 7.2 to 7.17 below, the surface linings of walls and ceilings should meet the following classifications:

TABLE 10
Classification of linings

Location	Class
Small rooms of area not more than: a. 4 m^2 in residential accommodation, and b. 30 m^2 in non-residential accommodation	3
Domestic garages of area not more than 40 m^2	3
Other rooms (including garages)	1
Circulation spaces within dwellings	1
Other circulation spaces, including the common areas of flats and maisonettes	0

Definition of walls

7.2   For the purposes of the performance of wall linings, a wall includes:

   (a) the surface of glazing (except glazing indoors), and

   (b) any part of a ceiling which slopes at an angle of more than 70° to the horizontal.

But a wall does not include:

(c) doors and door frames

(d) window frames and frames in which glazing is fitted

(e) architraves, cover moulds, picture rails, skirtings and similar narrow members, and

(f) fireplace surrounds, mantle shelves and fitted furniture.

Definition of ceilings

7.3   For the purposes of the performance of ceiling linings, a ceiling includes:

(a) the surface of glazing

(b) any part of a wall which slopes at an angle of 70° or less to the horizontal.

But a ceiling does not include:

(c) trap doors and their frames

(d) the frames of windows or rooflights (see Appendix E) and frames in which glazing is fitted

(e) architraves, cover moulds, picture rails and similar narrow members.

Variations and special provisions

Walls

7.4   Parts of walls in rooms may be of a lower class than specified in paragraph 7.1 (but not lower than Class 3) provided the total area of those parts in any one room does not exceed one half of the floor area of the room, subject to a maximum of 20 m² in a residential building and 60 m² in a non-residential building.

Fire protecting suspended ceilings

7.5   A suspended ceiling can contribute to the overall fire resistance of a floor/ceiling assembly. Such a ceiling should satisfy paragraph 7.1. If the assembly is to achieve 60 minutes fire resistance or more, it should also meet the provisions of Appendix A Table A3.

Fire-resisting ceilings

7.6   Cavity barriers are needed in some concealed floor or roof spaces (see Section 9); however, this need can be reduced by the use of a fire-resisting ceiling below the cavity. Such a ceiling should comply with Diagram 31, which stipulates a Class 0 surface on the soffit.

Rooflights

7.7   Rooflights should meet the relevant classification in 7.1. However, plastic rooflights with at least a Class 3 rating may be used where 7.1 calls for a higher standard, provided the limitations in Table 11 below and in Table 18 are observed.

Thermoplastic materials

General

7.11   Thermoplastic materials (see Appendix A, paragraph 16) which cannot meet the performance given in Table 10 can nevertheless be used in windows, rooflights and lighting diffusers in suspended ceilings if they comply with the provisions described in paragraphs 7.12 to 7.16 below. Flexible thermoplastic material may be used in panels to form a suspended ceiling if it complies with the guidance in paragraph 7.17. The classifications used in paragraphs 7.12 to 7.17, Table 11 and Diagram 24 are explained in Appendix A, paragraph 19.

Windows and internal glazing

7.12   External windows to rooms (though not to circulation spaces) may be glazed with thermoplastic materials, if the material can be classified as a TP(a) rigid product.

Internal glazing should meet the provisions in paragraph 7.1 above.

Notes:
1. A wall does not include glazing in a door (see paragraph 7.2);
2. Attention is drawn to the guidance on the safety of glazing in Approval Document N, Glazing - safety in relation to impact, opening and cleaning.

Rooflights

7.13   Rooflights to rooms and circulation spaces (with the exception of protected stairways) may be constructed of a thermoplastic material if:

(a) the lower surface has a TP(a) (rigid) or TP(b) classification

(b) the size and disposition of the rooflights accords with the limitations in Table 11 and with the guidance to B4 in Table 19.

DIAGRAM 23
Lighting diffuser in relation to ceiling
See paragraph 7.14.

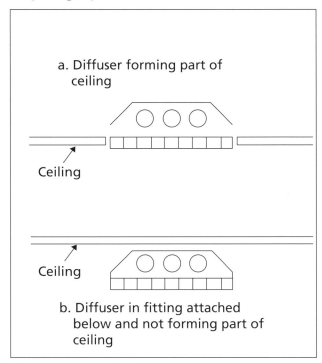

a. Diffuser forming part of ceiling

Ceiling

Ceiling

b. Diffuser in fitting attached below and not forming part of ceiling

Lighting diffusers

7.14   The following provisions apply to lighting diffusers which form part of a ceiling, and are not concerned with diffusers of luminaires which are attached to the soffit, or suspended beneath a ceiling (see Diagram 23).

Lighting diffusers are translucent or open structured elements that allow light to pass through. They may be part of a luminaire or used below rooflights or other sources of light.

7.15   Thermoplastic lighting diffusers should not be used in fire-protecting or fire-resisting ceilings, unless they have been satisfactorily tested as part of the ceiling assembly that is to be used to provide the appropriate fire protection.

7.16   Subject to the above paragraphs, ceilings to rooms and circulation spaces (but not protected stairways) may incorporate thermoplastic lighting diffusers if the following provisions are observed:

(a) wall and ceiling surfaces exposed within the space above the suspended ceiling (other than the upper surfaces of the thermoplastic panels) should comply with the general provisions of paragraph 7.1 according to the type of space below the suspended ceiling

(b) if the diffusers are of classification TP(a) (rigid), there are no restrictions on their extent

(c) if the diffusers are of classification TP(b), they should be limited in extent as indicated in Table 11 and Diagram 24.

Suspended or stretch skin ceilings

7.17   The ceiling of a room may be constructed either as a suspended or stretched skin membrane from panels of a thermosetting material of the TP(a) flexible classification, provided that it is not part of a fire-resisting ceiling. Each panel should not exceed 5 m² in area and should be supported on all its sides.

DIAGRAM 24
Layout restrictions on Class 3 plastic rooflights, TP(b) rooflights and TP(b) lighting diffusers

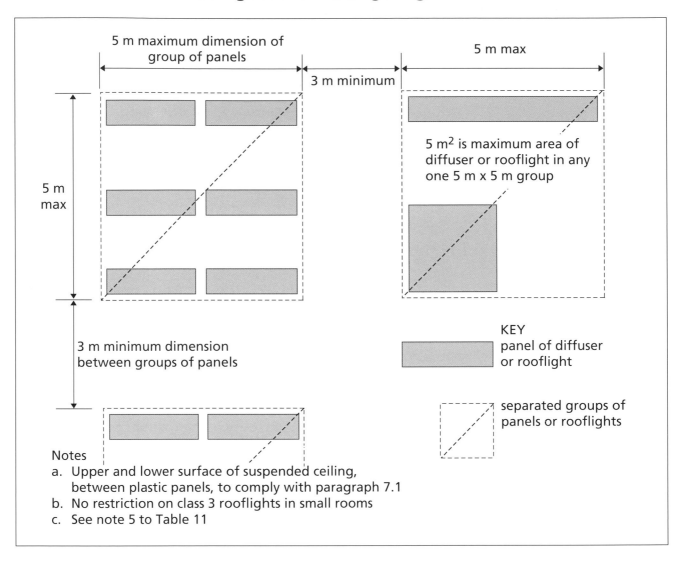

5 m maximum dimension of group of panels

3 m minimum

5 m max

5 m max

5 m² is maximum area of diffuser or rooflight in any one 5 m x 5 m group

3 m minimum dimension between groups of panels

KEY

panel of diffuser or rooflight

separated groups of panels or rooflights

Notes
a. Upper and lower surface of suspended ceiling, between plastic panels, to comply with paragraph 7.1
b. No restriction on class 3 rooflights in small rooms
c. See note 5 to Table 11

TABLE 11
Limitations applied to thermoplastic rooflights and lighting diffusers in suspended ceiling and Class 3 plastic rooflights

Minimum classification of lower surface	Use of space below the diffusers or rooflight	Maximum area of each diffuser panel or rooflight(1)	Max total area of diffuser panels and rooflights as percentage of floor area of the space in which the ceiling is located	Minimum separation distance between diffuser panels or rooflights(1)
TP(a)	any except protected stairway	No limit[2]	No limit	No limit
Class 3[3] or TP(b)	rooms	5 m^2	50 %[4][5]	3 m[5]
	circulation spaces except protected stairways	5 m^2	15 %[4]	3 m

Notes:

(1) Smaller panels can be grouped together provided that the overall size of the group and the space between one group and any others satisfies the dimensions shown in Diagram 24.

(2) Lighting diffusers of TP(a) flexible rating should be restricted to panels of not more than 5 m^2 each, see paragraph 7.17

(3) There are no limitations on Class 3 material in small rooms.

(4) The minimum 3 m separation specified in Diagram 24 between each 5 m^2 must be maintained. Therefore, in some cases it may not be possible to use the maximum percentage quoted.

(5) Class 3 rooflights to rooms in industrial and other non-residential purpose groups may be spaced 1800 mm apart provided the rooflights are evenly distributed and do not exceed 20 % of the area of the room.

Notes to Appendix B

Thermoplastic materials

(1) A thermoplastic material means any synthetic polymeric material which has a softening point below 200 °C if tested to BS 2782 Part 1: Method 120A: 1976. Specimens for this test may be fabricated from the original polymer where the thickness of material of the end product is less than 2.5 mm.

(2) A thermoplastic material in isolation cannot be assumed to protect a substrate, when used as a lining to a wall or ceiling. The surface rating of both products must therefore meet the required classification. If, however, the thermoplastic material is fully bonded to a non-thermoplastic substrate, then only the surface rating of the composite will need to comply.

(3) Concessions are made for thermoplastic materials used for windows, rooflights and lighting diffusers within suspended ceilings, which may not comply with the criteria specified in paragraphs A10 et seq. They are described in the guidance on requirements B2 and B4.

(4) For the purposes of the requirements B2 and B4 thermoplastic materials should either be used according to their classification 0-3, under the BS 476 Parts 6 and 7 tests as described in paragraphs A10 et seq., if they have such a rating, or they may be classified TP(a) rigid, TP(a) flexible, or TP(b) according to the following methods:

TP(a) rigid:

(i) rigid solid pvc sheet

(ii) solid (as distinct from double or multiple skin) polycarbonate sheet at least 3 mm thick

(iii) multi-skinned rigid sheet made from unplasticised pvc or polycarbonate which has Class I rating when tested to BS 476 Part 7 : 1971 or 1987

(iv) any other rigid thermoplastic product, a specimen of which, when tested to BS 2782 : 1970 as amended in 1974: Method 508A, performs so that the test flame extinguishes before the first mark, and the duration of flaming or afterglow does not exceed 5 seconds following removal of the burner.

TP(a) flexible:

Flexible products not more than 1 mm thick which comply with the Type C requirements of BS 5867 Part 2 when tested to BS 5438, Test 2 1989 with the flame applied to the surface of the specimens for 5, 15, 20 and 30 seconds respectively, but excluding the cleansing procedure.

TP(b):

(i) rigid solid polycarbonate sheet products less than 3 mm thick, or multiple skin polycarbonate sheet products which do not qualify as TP(a) by test; or

(ii) other products which, when a specimen of the material between 1.5 and 3 mm thick is tested in accordance with BS 2782 : 1970, as amended in 1974: Method 508A, has a rate of burning which does not exceed 50 mm/minute. (If it is not possible to cut or machine a 3 mm thick specimen from the product then a 3 mm test specimen can be moulded from the same material as that used for the manufacture of the product).

Appendix C: Building Standards (Scotland) Regulations 1990

as amended by the Building Standards (Scotland) Amendment Regulations 1993, the Building Standards (Scotland) Amendment Regulations 1994, the Building Standards (Scotland) Amendment Regulations 1996, the Building (Scotland) Amendment Regulations 1997, the Building Standards and Procedure Amendment (Scotland) Regulations 1999, and the Building Standards Amendment (Scotland) Regulations 2001.

Extracts from the Building Standards (Scotland) Regulations 1990 as amended that have particular relevance to this Guidance Note are included below. Persons working in Scotland will need all the relevant Technical Standards.

Compliance with building standards

9. (1) The requirements of Regulations 10 to 32 shall be satisfied <u>only</u> by compliance with the *relevant standards.*

   (2) Without prejudice to any other method of complying with a *relevant standard*, conformity with provisions which are stated in the *Technical Standards* to be deemed to satisfy that standard shall constitute such compliance.

Electrical installations

26. (1) Every electrical installation to which this Regulation applies and every item of stationary electrical equipment connected to such an installation shall provide adequate protection against its being a source of fire or a cause of personal injury.

   (2) This Regulation shall not apply to an installation:

      (a) serving a *building* or any part of a *building* to which the Mines and Quarries Act 1954(a) or the Factories Act 1961 applies;

(b) forming part of the works of an undertaker to which Regulations for the supply and distribution of electricity made under the Electricity (Supply) Acts 1882 to 1936 or Section 16 of the Energy Act 1983(b) apply; or

(c) consisting of a circuit (including a circuit for telecommunication or for transmission of sound, vision or data, or for alarm purposes) which operates at a voltage not normally exceeding 50 V a.c. or 120 V d.c., measured between any two conductors or between any conductor and earth, and which is not connected directly or indirectly to an electricity supply which operates at a voltage higher than those mentioned in this sub-paragraph.

(3) In paragraph (1) 'stationary electrical equipment' means electrical equipment which is fixed, or which has a mass exceeding 18 kg and is not provided with a carrying handle.

Access and facilities for *dwellings*

29.(1) A *building* of *purpose group* 1 shall be provided with:

(a) safe and convenient access from a suitable road;

(b) adequate access between its *storeys*;

(c) adequate sleeping accommodation;

(d) adequate *kitchen* facilities;

(e) adequate windows;

(f) adequate space heating.

(2) Every *building* of *purpose group* 1 to which it is *reasonably practicable* to make available a public supply of electricity shall be provided with sufficient electricity lighting points and *socket-outlets*.

(3) This Regulation shall not be subject to specification in a notice served under Section 11 of *the Act*.

Lighting of escape routes and circulation areas

E9.1 Every part of an *escape route* must have artificial lighting providing a level of illumination not less than that provided by suitable *emergency lighting* supplied with electric current -

   a. by a *protected circuit*; and

   b. where it serves a *protected zone*, by a separate circuit from that supplying any other part of the *escape route*,
   except -
   a *protected circuit* is not required where *emergency lighting* is installed in accordance with E9.2,
   except -
   in a *building* to which Part 1 of the Cinematographic (Safety) (Scotland) Regulations 1955 apply.

Emergency lighting

(E9.2) The requirements of E9.2 will be met where *emergency lighting* is installed in -

   a. cinemas, bingo halls, ballrooms, dance halls and bowling alleys, in accordance with CP1007: 1955; and

   b. any other *building*, in accordance with BS 5266: Part 1: 1999.

In the case of a *building* with a smoke and heat exhaust ventilation system the emergency lighting should be installed so that it is not rendered ineffective by smoke filled reservoirs.

TABLE E9.2
Emergency lighting

Purpose group or sub-group	Part of a building requiring emergency lighting
1A other than a *dwelling*, 2-7	1. A *protected zone* or *unprotected zone* in a *building* with a *storey* at a height of more than 18 m.
	2. A *room* with an *occupancy capacity* of 60 or more or any *room* containing an *inner room* with an *occupancy capacity* of more than 60, and any *protected zone* or *unprotected zone* serving such a room.
	3. An underground car park including any *protected zone* or *unprotected zone* serving it where less than 30% of the perimeter of the car park is open to the external air.
	4. A *protected zone* or *unprotected zone* serving a *basement storey*.
	5. A *place of special fire risk* (other than one requiring access only for the purposes of maintenance) and any *protected zone* or *unprotected zone* serving it.
	6. Any part of an *air supported structure*, other than one ancillary to a *dwelling*.

Purpose group or sub-group	Requirements additional to 1-6 above
2 (other than a *hospital*)	7. A *room* with an *occupancy capacity* of more than 10 and any *protected zone* or *unprotected zone* serving such a room.
	8. A *protected* zone or *unprotected zone* serving a storey required to have 2 or more *escape routes* other than, subject to 1. above, a *storey* in a *building* not more than 2 *storeys* high with a floor area of not more than 300 m².
	9. A *protected zone* or *unprotected zone* in a single stair *building* of 2 *storeys* or more with an *occupancy capacity* of 10 or more.
2A *hospital*	10. Essential lighting circuits must be provided throughout and designed to provide not less than 30 per cent of the normal lighting level. [1, 2].
4	11. In shop premises, a *protected zone* or *unprotected zone* serving a *storey* required to have 2 or more *escape routes*.
	12. In an enclosed shopping centre with a mall on 2 or more *storeys* or having a total floor area more than 5,600 m², the mall and any *protected zones* or *unprotected zones* required to have at least 2 *escape routes*
5	13. A *protected zone* or *unprotected zone* serving - a. a *storey* required to have at least 2 *escape routes*; or b. any *storey* in a non-residential school of more than one *storey*.
6	14. A *protected zone* or *unprotected zone* serving a *storey* required to have at least 2 *escape routes*.
7A	15. A *protected zone* or *unprotected zone* serving a *storey* required to have at least 2 *escape routes*, other than in a *single-storey building* with a floor area of not more than 500 m².
7C	16. A *protected zone* or *unprotected zone* serving any *storey*.

Automatic detection in dwellings

(E11.1) The requirements of E11.1a will be met by a *smoke alarm* with a standby supply, complying with BS 5446: Part 1: 1990 and installed in accordance with the provisions of a. to e. below -

a. The standby power supply for the *smoke alarm* should take the form of a primary battery, a secondary battery or a capacitor.

The capacity of the standby supply should be sufficient to power the *smoke alarm* when the mains power supply is off for at least 72 hours while giving an audible warning of mains power supply being off. There should remain sufficient capacity to provide a warning of smoke for a further 4 minutes. An audible warning should be given at least once every minute where the capacity of the standby power supply falls below that required to satisfy the recommended standby duration when the mains power supply is on; or persist for at least 15 days when the mains power supply is off.

b. A *smoke alarm* should be located -

   i. in a *circulation area* which will be used as a route along which to escape, not more than 7 m from the door to a living *room* or *kitchen* and not more than 3 m from the door to a room intended to be used as sleeping accommodation, the dimensions to be measured horizontally,

   ii. where the *circulation area* is more than 15 m long, not more than 15 m from another *smoke alarm* on the same *storey*,

   iii. where designed for ceiling mounting, at least 300 mm away from any wall or light fitting, or if designed for wall mounting, more than 150 mm and not more than 300 mm below the ceiling,

   iv. at least 300 mm away from, and not directly above, a heater or air conditioning outlet, and

   v. on a surface which is normally at the ambient temperature of the rest of the *room* or *circulation area* in which the *smoke alarm* is situated.

   Note:
   The above provisions are broadly in line with the recommendations of BS 5839: Part 6: 1995 for a Grade D Type LD3 system.

c. Where more than 1 *smoke alarm* is installed in a *dwelling* they should be interconnected so that detection of a fire by any one of them operates the alarm signal in all of them.

   d. A *smoke alarm* should be permanently wired to a circuit. The mains supply to the *smoke alarm* should take the form of either -

   i. an independent circuit at the *dwelling's* main distribution board, in which case no other electrical equipment should be connected to this circuit (other than a dedicated monitoring device installed to indicate failure of the mains supply to the *smoke alarms*), or

   ii. a separately electrically protected, regularly used local lighting circuit.

   Note:
   Where *smoke alarms* are of a type that may be interconnected, all *smoke alarms* should be connected on a single final circuit.

   e. Any *smoke alarm* in a *dwelling* which forms part of residential accommodation with a warden or supervisor, should have a connection to a central monitoring unit so that in the event of fire the warden or supervisor can identify the *dwelling* concerned, and the system should comply with the recommendations in BS 5839: Part 6: 1995 for a Grade C Type LD3 installation.

The requirements of E11.1b will be met by an automatic fire detection and alarm system complying with BS 5839: Part 1: 1988: Type L3.

Note:
BS 5839, Part 1 : 1988 was replaced by BS 5839-1 : 2002 on the 15th July 2003. The categories (previously 'types') of system such as L1, L2 and L3 have been extended in number in the new code of practice to L1, L2, L3, L4 and L5. Persons involved in fire alarm systems will need to consult BS 5839-1 : 2002.

BS 5839-1

Copies of the Building Standards (Scotland) Regulations and Scottish Office Technical Standards may be obtained from:
The Stationery Office Bookshop, 71 Lothian Road, Edinburgh, EH3 9AZ.
Telephone orders: 0870 606 5566.
Fax orders: 0870 606 5588.

Index

NOTES

NOTES

NOTES

NOTES

NOTES

NOTES